THE ANIMAL
AND
THE THINKER

THE ANIMAL
AND
THE THINKER

Instinct, Reason and the
Dance of Our Divided Selves

John Duncan

WH ALLEN

UK | USA | Canada | Ireland | Australia
India | New Zealand | South Africa

WH Allen is part of the Penguin Random House group of companies
whose addresses can be found at global.penguinrandomhouse.com

Penguin Random House UK
One Embassy Gardens, 8 Viaduct Gardens, London SW11 7BW

penguin.co.uk
global.penguinrandomhouse.com

First published by WH Allen in 2025
1

Gender & Culture by Melford E. Spiro is quoted in this book
by kind permission of Taylor and Francis Group
Populism by Cas Mudde and Cristóbal Rovira Kaltwasser is quoted in this book
by kind permission of Oxford Publishing Limited (Academic)
On Aggression by Konrad Lorenz is quoted in this book
by kind permission of Taylor and Francis Group
Human Ethology by Irenäus Eibl-Eibesfeldt is quoted in this book
by kind permission of Taykor and Francis Group
Love & Hate by Irenäus Eibl-Eibesfeldt is quoted in this book
by kind permission of Bernholf Eibl-Eibesfeldt

Typeset in 11.7/16pt Calluna by Jouve (UK), Milton Keynes
Printed and bound in India by Manipal Technologies Limited

The authorised representative in the EEA is Penguin Random House Ireland,
Morrison Chambers, 32 Nassau Street, Dublin D02 YH68

A CIP catalogue record for this book is available from the British Library

HARDBACK ISBN 9780753560921
TRADE PAPERBACK ISBN 9780753560938

Penguin Random House is committed to a sustainable future
for our business, our readers and our planet. This book is
made from Forest Stewardship Council® certified paper.

To my friends and colleagues in science
for 50 years of keeping me on my toes

my family on the farm
for showing me the many sides of life

and my family in Cambridge
for everything

Contents

CONTENTS

PART V
Power

Introduction
Pan and Spock

Who are we?

Along the rim of a leaf on a sallow bush sits the caterpillar of an eyed hawkmoth. Its shape and colouring make it hard to spot in this position on the leaf. Its only real tasks are to stay in a safe place and to eat, though periodically there is something new: its hard skin is now too small for its expanding bulk, and it goes through a crisis as its skin splits, it wriggles free, and a new, larger skin is ready underneath. The life of the caterpillar is very simple, and it is run by a correspondingly simple nervous system. As far as I know, nobody has counted the neurons in a hawkmoth caterpillar but, based on what we know of other insects, it will be in the range of tens to hundreds of thousands.

In her living room sits a teenage girl doing her physics homework. Her mother comes in from work and says hello; the girl glances up and smiles but, honestly, right now she just wants to get on with the physics. She finds this part of her school-work fascinating, and over the past few months, she has come to realise this is not just intellectual engagement; she *loves* physics, and, peculiarly, though she is also good at French, she despises learning French vocabulary. Of course, she does not just love physics. Periodically her eyes stray to her phone, wondering what her friends will say about the boy who was looking at her in class. She is hoping that her little brother will not materialise to annoy her, and also hoping that he was not picked on again at school. From the next room she hears the TV news, perhaps about wild-fires, perhaps a Congressional hearing, perhaps war in Ukraine. Right now she would really like to ignore all this and stay with the clean world of her equations. In her head are around a hundred billion neurons, and as Meryl Streep and Alec Baldwin told us in the 2009 film, *It's Complicated*.

Any glance at the news shows just how complicated it is. Here are a few headlines:

'My Southport shop was looted by rioters, then saved by strangers.' In Southport, UK, Chanaka Balasuriya tells how his grocery store was destroyed by far-right rioters – then put back together by volunteers from his local community.

'Principal resigns after Florida students shown Michelangelo statue.' Parents in Florida complained that an art class showing Michelangelo's *David* was pornographic. Principal Hope Carrasquilla resigned after being given an ultimatum to resign or be fired.

'Met Police: Women and children failed by "boys' club", review finds.' A 'blistering review' found racism, misogyny and homo-phobia to be rife in London's police force.

'Man City v Liverpool: I should become more selfish – Jack Grealish.' Transferred to the fabulously wealthy Manchester City for £100m, the football star explains his first-day nerves, fearing judgement from his new team-mates.

'Ukraine war: "My city's being shelled, but mum won't believe me".' Oleksandra, a Russian woman living in Ukraine, explains how hard it is to speak to her mother back home, who refuses to believe that her daughter is in danger.

'Army officer completes remarkable solo South Pole trek.' Preet Chandi completes her solo trek to the South Pole. She says she wanted to do it 'for people who don't fit a certain image'.[1]

Or you can glance over the comments posted under songs on YouTube. Under Tom Petty's 1989 song 'Running Down a Dream' can be found this piece of high art, in 68 words capturing love – shared ritual – death – loss – immortality – mystery – the power of music:

> My husband and I would play this song, every Friday night before we went out to our favorite beach bar. He died this week. The day after, I was standing in line at the Target pharmacy, waiting to pick up a prescription. This song came on. I don't remember ever hearing music in Target before. I think this was his way of contacting me. We were married 58 years.[2]

Often, in stories like these, it seems human beings make no sense – our rationality and our irrationality, our generosity and our greed, our dissatisfaction and our joy, our ambition and our laziness, our good and our evil, our brilliance and our blinkered half-truths, our ideals and our compromises. Throughout history, thinkers, writers and philosophers have struggled with the great questions of human life. Head and heart. Right and wrong. Freedom and responsibility. Women and men. Democracy and justice. Ambition and fulfilment. The questions are everywhere – in our aspirations, our institutions, our burning beliefs and burning conflicts, our search for a sense of meaning. Who really are we, what do we need and how do our lives work?

Answers require knowledge – in this case, knowledge of ourselves. The knowledge we need, I shall argue, comes from a balance between two sides of ourselves. The two sides have received a good deal of scientific attention but are rarely put together. On the one hand, we need the science of the *reasoning human mind*. The science here concerns information processing, and the idea of the brain as a staggeringly powerful, general purpose computer. On the other hand, we need the fundamental laws of *animal behaviour*, established by the great ethologists of the twentieth century. To understand and manage our lives, we need all the details of both sides – the strengths and weaknesses of the mind as an idea machine, and the elaborate needs and behaviour patterns of the complex human animal.

Reason is not enough

At least since the ancient Greeks, it has been felt that humans should guide their lives by the force of reason. In the fourth century BC, Plato considered reason to be the monarch of the psyche, ruling over its other parts. In the thirteenth century, Thomas Aquinas considered that human affairs were governed by natural laws, to be discovered by reason. It is perhaps natural to feel that the power of reason is what sets human beings apart, allowing us, unlike all other animals, to base our thoughts, our choices and our social institutions on understanding and truth. The power of human reason is certainly staggering. As we look around our modern world – cities of concrete and glass, rice fields, microchips, transport networks, communications satellites – it is our own reasoning minds that have built it. Commonly, too, we feel that reason is the key to moral choices and to how we should live our lives. The sense is that somewhere out there are essential laws of right and wrong – an external reality of good and bad choices. When we violate abstract principles of justice, fairness or human rights, or live a trivial and lazy life, we are violating not just human laws, but laws of the universe.

With the establishment of psychology as an experimental science, the mechanics of human reason have come under increasing scrutiny. In the 1940s there was the work of a great German psychologist Karl Duncker, examining detailed protocols of the thought process as his participants struggled to work out how one should direct X-rays to destroy a tumour without harming the surrounding tissue.[3] By the 1950s we had the first digital computers, and very much in the spirit of Duncker, the pioneers Allen Newell and Herbert Simon wrote the first problem-solving programs, proving theorems in formal logic in a way strongly reminiscent of human problem solvers.[4]

I have spent much of my own research life pulling together clues to the nature of human reason from both classical experimental psychology and the rapidly developing field of human neuroscience. There are simple reasoning tests that are often used as measures of 'intelligence'. Why is it that for some people these tests are trivial, for other people impossible? There are clues from the study of brain damage following stroke and other diseases, with spectacular and often bizarre reasoning deficits produced by damage to the frontal lobes of the brain. In the 1990s 'cognitive neuroscience' was born with the opportunity to measure human brain activity using MRI machines, and soon we were using this new technology to investigate brain activity during classical reasoning tests. There is neurophysiology, with recordings of the electrical activity from large populations of neurons, each carrying its own message in the detailed pattern of electrical impulses it sends out. Of course, there is a great deal that is still under debate, but in 50 years, things have moved very far forward. As I explained in a previous book, *How Intelligence Happens* (2010), I think that in 2025 we have a good outline map of human reason and the brain mechanisms that produce it.

With its flexibility and its power, the human reason machine is certainly fascinating, and it is certainly one pinnacle of the human mind. But if we seek answers to our big human questions, we see that reason alone cannot be enough. It cannot tell us why our hearts melt at a smile from our child, or swell with pride when our country wins gold at the Olympics. I will argue that, taken on its own, reason cannot tell us what is right and what is wrong, what laws we should pass or how we should run our society. Above all, reason cannot tell us 'the meaning of life'. To explain these things, we need reason plus something more.

Another side

I took an erratic route into psychology. Entering my final year of school, I thought I would be a biologist. My childhood on an English dairy farm had been filled with animals and plants. Thinking that the children should learn how to make their own money, my father put us in charge of rearing pigs; in the mornings, we rushed out to feed them before breakfast, often with slightly disastrous results for the state of my trousers on the school bus. I was fascinated by insects; aged about six, I learned about ants with a surprising pain when I turned over the wrong stone on a disused railway line, and about the orange-tip butterfly by its first arrival in spring in a meadow by the river. With my grandfather, I took early morning walks to count the heifers; I learned how far ahead of the heifer's thinking you need to be to persuade it through a gate and, meanwhile, my grandfather showed me plants in the hedgerows and taught me some botany.

At school, I much preferred biology to the physical sciences, with the chance to lessen the tedium of homework with the interest of living things. In A-level biology, for example, we studied the question of how water travels up from the roots to the crown of a tree. The trunk is filled with tiny tubes called xylem, but how can the water be lifted up such a long tube? I was taken with a rather elegant experiment in which one saw cut was made three-quarters of the way through a tree trunk at one level, then one higher up, three-quarters of the way through from the other side; there could be no one continuous tube, but the tree still lived. The science was interesting to me, and so was the sawing; helping my father cut firewood was something rather precious to me. I thought this could be my future, and applied to university to study botany.

One evening, however, I was visiting my grandmother, who owned a TV, and we watched a 1940s noir which I have subsequently tracked down as *The Dark Past*. An escaped convict breaks into a psychiatrist's home and holds the family hostage. Outside, there is a police siege; inside, the psychiatrist slowly uncovers the forgotten childhood that turned his 'patient' to the dark side. Suddenly this seemed even more interesting than xylem, and I promptly changed my application to psychology and physiology, the closest thing that Oxford University offered to a degree in psychiatry.

Of course, at this point I wanted to be a psychiatrist, but soon afterwards I discovered there was more to the mind than *The Dark Past*. Now that I was going to study psychology, a school friend lent me his copy of Konrad Lorenz's *On Aggression* – written in 1963 and translated into English in 1966 – and told me that nowadays psychology was not all about serial killers and Freud. Lorenz was one of the fathers of ethology, the science of animal behaviour as a branch of biology, and in his hands, insights into the principles of instinct were used to illuminate deep questions of human life.

I fell into the pages of his book with that thrilled sense of new understanding that is one of the joys of the growing human mind. In the first chapter, Lorenz is diving off the Florida Keys, observing the fighting displays of coral fish, and I found this so spellbinding that, many years later, I finally put to one side a quite significant reluctance to inhale underwater and learned to dive over the reef myself. Unlike Lorenz, I was rewarded only with a child's fascination at appearing to swim in an aquarium, with clouds of sparkling fish like swarms of butterflies, rising from the blue depths and swirling up over the side of the multicoloured reef. In the hands of Lorenz, however, this fascination was turned into something different. In *On Aggression* there are battling stags, hordes of rats held together by a common odour, squabbling ducks and geese, fish and birds of all kinds, a teeming world distilled into simple principles of the instinctive animal mind at work. At the same time, against this teeming background, the book showed how inspired observation of animal behaviour can lead to unexpected, sometimes uncomfortable but compelling insights into ourselves.

Later, I was to discover that Lorenz had also been a Nazi, who had worked as a psychologist with the SS before finally being dispatched to the Russian front. By the time he wrote *On Aggression*, of course, Lorenz had long renounced his early sympathies. In a later chapter I will come back to the discord we feel when we know that Lorenz the thinker had also, at one time, been Lorenz the Nazi. This discord notwithstanding, the core ideas of Lorenz and his ethological colleagues are essential, I shall argue, to understanding both animal and human minds.

Of course, when I actually reached university, I discovered that the science of the mind has many sides and, as I have said, my life in research has moved far away from this early interest in ethology. This is not to say that the core ideas of *On Aggression* have lost their force; indeed, if I now turn back to the book after

50 years, I am simply astonished at how much it has continued to mark my views of the human condition – some well-remembered, but some that I have cherished for many decades as my own original thoughts, only now to discover that I first read them in Lorenz, forgot them, then later 'made them up' as my own! (In *On Aggression*, for example, Lorenz the erstwhile Nazi puts forth deep ideas on group conflict and its threat to peace. As this book progresses, we shall see that consistency is not at all a reliable principle of the human mind.)

The same applies to another book that I read as an under-graduate, Irenäus Eibl-Eibesfeldt's *Love and Hate* (first published in German in 1970). Eibl-Eibesfeldt had learned from Lorenz and became a long-term colleague, but where Lorenz simply loved watching animals, then generalised his conclusions to the human, Eibl-Eibesfeldt took the human case head on, applying etholog-ical methods and principles to detailed, cross-cultural data on human social behaviour. Once again, looking back at *Love and Hate*, I am struck by how firmly, 50 years ago, its ideas took root in my mind.

Now, in the current book, I should like to put two things together – *the science of human reason*, which has occupied so much of my own research life, and *the principles of animal behav-iour*, as developed by Lorenz, Eibl-Eibesfeldt and other giants of ethology. To understand ourselves, I shall argue, we need to know how both sides of our minds work; and when we understand ourselves, we understand much of the puzzle and uncertainty of human life.

Perhaps it is no surprise that, in broad terms, this balance of animal and idea was understood already by Charles Darwin. Here he is in *The Descent of Man* (1871), discussing the merge of instinct and reason in human morality:

the social instincts which no doubt were acquired by man, as by the lower animals, for the good of the community, will from the first have given to him some wish to aid his fellows, and some feeling of sympathy. Such impulses will have served him at a very early period as a rude rule of right and wrong. But as man gradually advanced in intellectual power and was enabled to trace the more remote consequences of his actions; as he acquired sufficient knowledge to reject baneful customs and superstitions; as he regarded more and more not only the welfare but the happiness of his fellow-men; as from habit, following on beneficial experience, instruction, and example, his sympathies became more tender and widely diffused, so as to extend to the men of all races, to the imbecile, the maimed, and other useless members of society, and finally to the lower animals – so would the standard of his morality rise higher and higher.[5]

Pan and Spock

In Greek mythology, Pan was the god of nature, fertility and the wild. Pan has the body and head of a man, but the horns, legs and hooves of a goat. Pan is the animal side of ourselves. He loves music. He is the companion of the nymphs.

In 1966, when *Star Trek* and the starship *Enterprise* first arrived on TV screens, viewers were introduced to the pointy-eared Spock. Spock, half Vulcan, has the perfect rational mind. He appears to know everything, and when Captain Kirk asks his advice, the answer is complete, balanced and precise. But there is something missing in Spock. Spock does not get humour. Spock would never pump his fist in triumph. Spock's face is composed. Spock is perfectly rational, but he is only half human.

In Pan and Spock we have human instinct (the 'animal' in us) and human reason (the 'idea[s]' we create), each with their own urging on how to live our lives, and each with their own essential contribution to our humanity. Both, of course, have been built by millions of years of evolution, but their operating principles are quite different. Spock – our reasoning side – is an elaborate computing machine, marshalling immense banks of knowledge to solve a countless multitude of novel, constantly varying problems. Evolution built the machine, but learning fills it with knowledge of toys, shops, theorems, chess, the big bang – all the knowledge it needs to tell us what will work and what will not as we navigate our way through our modern lives. In contrast to the flexibility of Spock, Pan – our instinctive side – is a heavily innate, ethological set of fixed action patterns and modes of thought, similar in general principle to the instincts of animals from insects to great apes. Across the animal kingdom, complex instincts control how the members of a species interact – in competing for resources, in attracting mates, in caring for the young, in defending the group. We are the same, and in our own, highly social species, Pan contributes elaborate procedures controlling our interactions with the people around us.

Both Spock and Pan are immensely powerful – but both also have profound weaknesses and blind spots. These two voices in our heads live in uneasy truce; sometimes they recommend the same course of action, but very often they are in screaming conflict. Together, they have led humanity to its heights and its crises. It is unwise to ignore either one. It is equally unwise to believe either one. To understand and manage our lives, I shall argue, we need both.

There is an implication in all this. We may think of ourselves as coherent, reasonable people, but this is not what our minds are like. Each person is a patchwork of many things – complex, competing instincts calling out to be satisfied, ideas that change from one moment to the next, sometimes in agreement, often in conflict. The unified 'self' is an illusion. When we understand the operating principles of instinct and reason, we see that there is no self beyond the patchwork. We are creatures of conflict – conflict with others, and conflict with ourselves.

As Spock and Pan work together, furthermore, they have more to struggle with than one another. As they agree and disagree, both are dealing with a world for which neither was prepared. Evolution shaped them, but not evolution in a world like ours. Up to around 10,000 years ago, *Homo sapiens* lived in small communities of hunter-gatherers. Our *social instincts* were shaped by the needs of these small communities and the social roles they contained. Our *reasoning power* was shaped by the problems likely to be encountered in the hunter-gatherer world. Now, we face human concerns at a scale of billions of individuals, and challenges far beyond those of hunter-gatherers. Spock and Pan struggle together to manage life in our modern, global society. In *On Aggression*, Lorenz put it like this:

> An unprejudiced observer from another planet, looking upon man as he is today, in his hand the atom bomb, the product of his intelligence, in his heart the aggression drive inherited from his anthropoid ancestors, which this same intelligence cannot control, would not prophesy long life for the species.[6]

There are certainly good reasons for this pessimism, but at the same time, the unprejudiced observer would be forced to raise an eyebrow at how well humanity has done so far. On the timescale of evolution, the rapid progression from separate hunter-gatherer communities to today's globalised society is something quite unique. Our lives are very far from perfect, but as Steven Pinker documents in *Enlightenment Now* (2018), in major ways they just keep getting better. The biologist from another planet would likely have said this could never happen. I think it can only happen because, despite their struggles, Spock and Pan find ways to work together.

The voice of Spock

Of course, like thinkers throughout history, I agree that Spock's power of reason must play a central part in our lives. The science of reason, however, shows that the human Spock is actually not at all like the flawless reasoner of *Star Trek*. Whatever we do with our minds, it is rarely this (a joke I got from my son many years ago):

Three logicians go into a bar. The barman comes over and asks, 'Pints all round?' The first logician says, 'I don't know.' The second one says, 'I don't know.' The third one says, 'Yes!'

With a little effort we can work out what the third logician concludes from the answers of the first two, but this is definitely not what we usually do when we go into a bar. More than a century after the stories were written, Sherlock Holmes still captures our imagination because he thinks so differently from the bemused, more human Inspector Lestrade. For millennia, humanity has presumed itself to be the pinnacle of reasoning, but as artificial intelligence grows ever more powerful, we increasingly realise how we have overestimated ourselves. Very likely, as artificial systems grow ever more able to integrate vast bodies of knowledge, we will have increasing access to real Spocks, more able than we could ever be to draw the best possible conclusions from the data fed in. I will come back much later to the possibilities for improved social, economic and political decisions – perhaps not here yet, but very obviously close. Meanwhile, we are stuck with our own, Lestrade version, and in our case, I shall argue, we see a machine that works brilliantly well in some ways, but is all too limited in others. Unlike Sherlock Holmes or the Spock of *Star Trek*, our real Spock, I shall argue, is simply a generator of ideas. Some of the ideas are excellent, some logical, some even brilliant, but many are just ideas. When reasoning about social issues, in particular, Spock is given to thinking in simple rules that capture just a part of complex, messy social reality and, often, these rules lead him to a very foolish place.

One mistake is to put reason on a pedestal, as the highest essence of our humanity, but it is equally wrong to denigrate reason's power. It is often felt that reason is a weak or even impotent sovereign, unable to control the impulses of the animal within. In his excellent 2012 book *The Righteous Mind*, the moral psychologist Jonathan Haidt argues that reason simply follows human instinct – instinct tells us what to think, and reason follows behind, providing a glib rationalisation. Often, certainly, we do use Spock to *rationalise* rather than reason, and Haidt is concerned with moral reasoning, where this may be especially salient. In our daily lives, however, it is obvious that we use the power of Spock in everything that we do, from planning our morning cup of coffee to deriving a proof in geometry. It is not hard to see why philosophers through the ages have seen reason as the sovereign of the mind and the essence of humanity. Indeed, in many ways it is – with reason we see truth in a way that no other animal can remotely approach.

The social lives of human beings, furthermore, are evidently not just a matter of instinct. Human life takes place in the rich context of different cultures, each with its own values, ideas, rituals and economy. Many common themes can be seen across cultures, showing the power of Pan. We need Spock, though, to explain how countless new ideas are attached to instinctive common themes; how one group follows Jesus and another Allah, both with the same burning certainty; or how one group can believe that women are incapable of rational thought and then, a short hundred years later, can demand equality of the sexes.

The voice of Pan

In *The Strange Case of Dr Jekyll and Mr Hyde* (1886), Robert Louis Stevenson depicts the tension between the rational, civilised human being – the person we truly aspire to be – and the prowling, uncontrolled animal, knowing nothing but its hungers. In tales of the wolfman, the rational man is a creature of the light, but when the full moon rises, hair sprouts and the animal howls. We fear the animal but, in this book, I argue that the great thinkers of history got it wrong because, focused on human minds and lives, they did not give adequate attention to biology. The story told by Robert Louis Stevenson is only a part of the truth. In our animal side there is the pacing, violent Mr Hyde, but there is much more than this.

The Jekyll and Hyde story finds many adherents. In *The Chimp Paradox* (2012), Steve Peters argues that it is the thinker who is the 'real' human being. The thinker needs to suppress the chimp, a disruptive, often unwanted character who prompts only the simple animal urges of fight, flight, the search for food and the need for a mate. In *The Blank Slate* (2002), Steven Pinker argues that truly moral behaviour must have reasons for its choices; an animal urge is not a sufficient basis for a moral decision. Aristotle famously praised law as 'reason free from passion'. In all of this we see Dr Jekyll and Mr Hyde, and the human fear of our animal side.

I shall argue that many things are wrong with this view. Pan is not a simple creature at all, and if you read a little about animal behaviour, you soon see that, despite what I said about the eyed hawkmoth caterpillar, few creatures are that simple. The nervous system of a fish, a rat or even an insect equips it with an elaborate set of behaviour patterns, drawn out by the right eliciting conditions and allowing an equally complex, patterned life to unfold. If this is true of a social insect such as a honeybee, it is even more true of ourselves. As emphasised by evolutionary psychologists such as Leda Cosmides and John Tooby, we have evolved to navigate an astoundingly complex set of social relations, and in our Pan, we have an exquisitely crafted machine to guide us along the way.[7]

Pan is a sophisticated and complex guide, but his role in our lives is much more important than this. I shall argue that Pan is also essential, providing us with all that is deepest in our aspirations, our tastes, our sympathies and ourselves. In *A Treatise of Human Nature* (1739–40), the philosopher David Hume argued that reason on its own cannot possibly explain human behaviour. To motivate behaviour, reason needs a starting point, and to provide that starting point we need human desires, preferences and goals.

Suppose I wake up deciding whether to have eggs or pancakes for breakfast. Reason tells me that, if I want eggs, I get out the frying pan, but if I want pancakes, I get out the griddle. It does not tell me which one I feel like this morning, and this argument can always be pushed to a starting point that reason cannot possibly provide.

I find it useful to drive this point home with some extreme atrocity – I tend to use abuse of a child. Spock can tell the abuser, 'The child will be bitterly frightened and hurt.' The abuser replies, 'I don't care how the child feels.' Spock can say, 'The child will grow up distrustful, angry, alone, scarred.' The abuser replies,

'I don't care how the child grows up.' Spock can say, 'When children's lives are destroyed, the fabric of society is eroded.' The abuser replies, 'I don't care about the fabric of society.' Though moral philosophers struggle with this thought, in my view Hume is correct – ultimately, Spock has to be defeated, because whatever consequences he deduces, the abuser can refuse to recognise their significance. None of this matters to Pan, who is simply what evolution made him. Pan says, 'My friends and I value children, and if you harm them we will make you suffer.' There is no more he needs to say; he has given a complete, consistent account of himself and his intentions. Asking him why he thinks this way would be like asking a king penguin why he stands in the Antarctic blizzard protecting his egg on his feet. As Hume put it, 'a passion is an original existence', while 'reason is, and ought only to be, the slave of the passions'[8].

Of course, in a sense we all know this. At the end of the 2001 film *Legally Blonde*, having transformed from sorority airhead to head of her law class, Reese Witherspoon takes to task Aristotle and his characterisation of law as 'reason free from passion' – the law, she says, needs *both* the reason and the passion. A little shockingly, I find myself with *Legally Blonde* and against Aristotle.

Without Pan, we could never explain the essence of who we are and what laws we want. We could never understand why we are moved to offer food to a visitor, or why we throw ourselves into the sea to save an anonymous stranger. Equally, we could never understand our dark side, why Genghis Khan could proclaim, 'The greatest happiness is to vanquish your enemies, to chase them before you, to rob them of their wealth, to see those dear to them bathed in tears, to clasp to your bosom their wives and daughters.' Eibl-Eibesfeldt puts it this way: 'Our rational thought, initially developed as an instrument utilized for survival, has attained tremendous dimensions in the intellectual realm, but our archaic emotional side continues to be the center of our being.'[9]

And vitally, I argue, this animal side of ourselves is not just a curiosity, not just a biological remnant. Like other animals, we have instincts that are inconsistent, conflicting, dependent on time and circumstance, and deeply powerful. Like the instincts of other animals, they are built into our nervous systems with their own, spontaneous need to be discharged, and when this happens, they give a sense of rightness. In his masterpiece of thinking on the human condition, *Man's Search for Meaning*, first published in Austria in 1946, Viktor Frankl argued that we do not want happiness, we want meaning. It is the animal in us, I shall argue, that gives us the sense that our lives are not just an intellectual puzzle. For Pan, our lives have meaning.

The dialogue

It is a mistake to denigrate Pan and promote Spock. It is equally a mistake to denigrate Spock and promote Pan. Human life is a dialogue of these two essential forces in our minds and to manage our lives, we need all the complexity, all the strengths and all the weaknesses of both characters.

Spock and Pan – reason and instinct – work by different rules, and with their partial views of human nature, both can be unsettling. From the perspective of Spock, instinct can seem primitive and unsuited to the needs of civilised life. From the perspective of Pan, reason can seem dry and artificial, at variance with our true selves.

Reactions to *On Aggression* show the concerns of Spock. In his book, Lorenz argued that aggression between members of a species is extremely widespread in the animal kingdom, specifically in animals that compete for territory and mates; that it has a critical biological function; and that, like other competitive species, humans have strongly developed, instinctive aggressive behaviour and drives. As we shall see later, he went much farther than this, suggesting that bonds in many animals are cemented, perhaps even dependent on joint aggression to enemies. The conclusions of Lorenz were based on profound analyses of aggression and bonding in many species, from cichlid fish on the reef to a rich variety of birds and mammals, and few could sensibly doubt the strength of the principles he put forward. Still, when it came to humanity, many were unwilling to believe that just the same principles apply to ourselves. Typifying shocked reactions to his book, in the 'Seville Statement on Violence', written in 1986, UNESCO assembled a committee of 20 scientists to declare that Lorenz was wrong and that aggression is not built into human nature.

Human instincts – especially the instinct to aggression – often challenge us because they do not show human nature as we should like it to be. Unfortunately, desire is not fact, and I find it impossible to consider the aggressive displays of fish, birds, deer and monkeys, to survey the world around me, and to continue hoping that none of this applies to *Homo sapiens*. Saki puts this better than I ever could in his story 'The Toys of Peace' (published in 1919, three years after his death), in which an (approximately) Edwardian uncle tries to divert his young nephews from the joys of warfare with a collection of toys more morally uplifting than soldiers and dreadnoughts. The children's adaptation of their new toys is predictable but carried off with Saki's usual élan, the enacted slaughter of a hundred girls from the Young Women's Christian Association by the troops of Louis XIV being an especial high point. In the final sentence of the story, faced with the ruins of his re-education programme, the hapless uncle repines, 'We have begun too late.'[10] Indeed I think they have – by a few million years.

It is often argued that we build aggression into our children by the way we raise them, and certainly, we do not teach them always to lie down and die in the face of opposition. But surely the great majority of our experience as parents is directly the opposite of this. As our children change from babies to toddlers to schoolchildren, we spend immense effort instructing them that they must not bite their sister, or take their younger brother's toys, or band together with their friends to humiliate an outsider. In another of Saki's stories from *The Toys of Peace and Other Papers* collection, 'The Interlopers', two lifetime enemies are searching through a forest at night, each hoping that tonight at last they will be able to kill. Coming suddenly through the trees they meet, but there is a pause. As Saki puts it:

> Each had a rifle in his hand, each had hate in his heart and murder uppermost in his mind. The chance had come to give full play to the passions of a lifetime. But a man who has been brought up under the code of a restraining civilization cannot easily nerve himself to shoot down his neighbour in cold blood . . . [11]

Doubtless the influence of teaching and civilisation can go either way, but surely it restrains at least as often as it encourages the aggressions built into Pan.

Pan is not a saint, often far from it, but he is essential to who we are, and our lives would make no sense without him.

Meanwhile, as Spock struggles to accept the darker side of Pan, Pan can find Spock infuriating and irrelevant. I well remember how, as a child, I was told that Jesus thought we should always turn the other cheek, and I just felt, 'What absolute rubbish.' I thought this would not work in my playground, and almost nobody, I imagine, honestly believes that it is never right to stand up and fight. Spock has a bad habit of coming up with ideas that are too simple, and applying them too broadly. Pan, in contrast, is quite content to accept inconsistency; the schoolmate who is an ally on Monday can be happily punched on Tuesday. As we shall see in a later chapter, inconsistency is built deeply into animal behaviour.

While the essence of Pan is fixed in our genes, Spock's ideas evolve and spread within a single lifetime. Though new ideas often riff on one theme from Pan, often they conflict with another. Perhaps the most obvious example in current Western culture is the growth of political correctness and cancel culture. In line with the thinking of Lorenz – and as I shall be discussing in detail later – Pan uses a complex system of social forces to segregate human beings into us and them. Pan will die for 'us', but with almost no encouragement he is ready to turn violently on 'them'. Within my lifetime, the idea that we should speak of all people equally has seen a spectacular spread. It is a beautiful idea – but it conflicts violently with our 'them' mode, and when one person thinks 'us' and another 'them', we come into conflict not just with ourselves, but with one another.

In 1998, when my 13-year-old son told us he was gay, it was still something to be discussed and coped with – when to tell our parents? How to respond when colleagues casually mentioned 'playing for the other side'? Not many years later, the culture had changed, and disparagement had changed to acceptance. In a few short years my son could make me immensely happy by looking back and saying, 'I don't know what I was fussing about . . . it was just the gay thing.' The idea that we are all 'us' is overwhelmingly generous, and when we can feel it in that spirit, the rules of Spock are backed up by all the flooding warmth of Pan. Once, sitting in a park, I was recounting to a friend the plot of the 2001

Italian film *Le Fate Ignoranti* (The Ignorant Fairies) – in which a woman discovers that her now deceased husband had been having an affair with another man – and suddenly, overcome by the theme of acceptance, I found I was sobbing, uncontrollably, in public.

But conflict arises when Pan feels that the gays are 'them', and Spock tells him to keep quiet about it. I sometimes feel it is like telling a teenager to keep quiet. He may be quiet, but he is sulking in a corner waiting to get back at you when he can. Most parents are aware that there are better ways to deal with a teenager. In the case of Pan, as we shall be seeing, the operating principles of the nervous system mean that it cannot possibly work simply to tell Pan he is wrong – and in any case, actually, he is not. Despite the teachings of Jesus and our modern political correctness, our lives simply cannot do without the distinction of 'us' and 'them'. People are not all equal to us. We need to know and value who we are as individuals, who our friends are, what community we come from. The people around us are not abstract symbols of a common humanity – they are the concrete individual animals we know and love. Common humanity dictates that we should embrace all people but, at the same time, life cannot work without embracing the need to be ourselves.

When I had very small children, I was struck by how completely, in their mental world, desires could trump facts. The child is screaming for chocolate; you explain there is no chocolate in the car, but still the child continues to shout, 'Give me the chocolate!' My favourite example came many years ago as I left the building in Cambridge where I work. Just up the road is a nursery school, and in its playground, the trees are protected by metal grilles built around each trunk to perhaps four feet from the ground. As I walked by, I saw two small girls who had managed to throw their ball into the space behind one of these grilles; there it lay on the ground, thoroughly out of reach, and one girl was firmly instructing it: 'Ball! Ball! Come out, ball!'

As adults we may not shout for the ball but, still, it is surprisingly easy for us to believe in heavily over-simplified views of human nature, shaped not by experience but by Spock's cultural or political preferences. It is disturbing when our ideals are questioned, but always, with the tension of Spock and Pan, ideals capture only a part of complex reality.

What is to come

In this book I explain the conflicted world of Pan and Spock, of instinct and reason, of the animal and the thinker. Part I deals with operating principles. It explains the basic science of instinct and reason. In the remainder of the book, we turn to big questions – morality in Part II, ambition in Part III, sex in Part IV, politics in Part V. In each section we see how the basic operating principles of the mind illuminate the conflicts and needs of our half-civilised, half-innate human lives.

Chapter 1 begins with the principles of animal behaviour and the insights of the ethologists. Across the animal kingdom, instincts control behaviour from searching for food to avoiding predators to defending territory to attracting mates to caring for the young. They provide suitable solutions, simple or elaborate, to highly specific situations in the animal's life. They are often good on complexity and detail. They are weak on escaping from the detail of the moment to see a bigger picture. They are often conflicted, often incoherent and, as Lorenz clearly saw, often they have their own inner demand to be discharged.

Chapter 2 considers how these same principles apply to human instincts. Like a honeybee or a chimpanzee, we have our own, heavily innate patterns of behaviour and thought, from the facial expressions that communicate greetings, threats, feelings and thoughts in the finest detail, to the elaborate sequences of back-and-forth interplay between parent and child, to the specifically human constructs that we use to speak to, understand and imagine one another. Undoubtedly, the power of Spock gives us immense freedom to reimagine our own lives. At the same time, obviously enough, principles that have regulated animal behaviour throughout evolution do not suddenly stop with us.

In Chapter 3, I turn to the complementary strengths – and complementary weaknesses – of human reason. Reason, I shall argue, is a process of constructing new thoughts from essentially any components. This allows humans to create an infinite variety of social thoughts, beyond the imagination of any other animal – that women must cover their shoulders to approach the Wailing Wall, or that children should not speak until they are spoken to. An essential element in the world of Spock is abstraction – just a few things are selected and sewn together into a thought, while everything else is left out. This makes reason good for extracting rules that apply not just to one detailed situation, but to many situations with related properties. But the power of an abstraction is also its weakness. In an abstract rule, all the specific details of any one situation are left out. When the details matter, Spock tends to miss them – he believes his rule, and while Pan is screaming that Spock has it all wrong, Spock blindly continues to push the rule through. The results are everywhere, from the sterile rules of strict religious observance to the over-simplified culture wars of current Western society. When our thoughts are too simple, our humanity is diminished and our lives feel incomprehensible and wrong.

Like Pan, Spock is also troubled by his own conflicts. As he deliberates, attention is focused on just one part of the bigger picture, and as context shifts, so does Spock's focus. Pan calls out competing demands; Spock responds with conflicting ideas, each capturing just a part of complex reality. It's complicated.

Chapters 4 and 5 consider how Pan and Spock work together. In Chapter 4, I look briefly at the brain, and the different bases for Pan and Spock. Pan is based heavily in ancient brain structures, present throughout vertebrate evolution. Spock, in contrast, is something much more recent, best seen in images of the human cerebral cortex at work. Ancient and modern brain structures work together but, within them, Pan and Spock have very different representation. In Chapter 5, I turn to the spread of Spock's ideas. By much the same principles, I argue, ideas as well as animals evolve, but much faster, and with different mechanisms of generation and selection.

With the basic science in place, in the remainder of the book I turn to the big questions of human life. Part II, 'Good and Evil', deals with morality and its hidden but essential partner, friend and enemy. Pan, I shall suggest, brings the gushing sense of oneness when a crowd of a hundred thousand people rocks together to Queen, or when a child hugs a broken parent and says it will be OK. In Chapters 6 and 7, I deal with the bonding of human families and communities, and the instincts in place for mutual support, defence and care. But Pan also gave Genghis Khan the fierce joys of invasion and, today, produces the uncontrolled baying of a soccer crowd. In Chapter 8, I turn from 'us' to 'them', and the need for battle that runs just as deep as the need to love.

Meanwhile, the rules of Pan for hunter-gatherer societies simply cannot work in the large, urban communities that civilisation has created. To make life possible, we have the rules of Spock. Spock says, 'Thou shalt not kill,' and without this law, civilisation cannot work . . . but everybody knows that the real rule is ' . . .

except sometimes.' Simple laws can never capture the full complexity of our lives together, and the law is often an ass – but an ass we could not do without. Chapters 9 and 10 deal with the codification of moral principles, religion and legal systems. Blame, responsibility, free will . . . all these, I argue, are ideas that, like so many of Spock's over-simplifications, capture a part of reality but at the same time trap us in confusion.

In Part III, 'Ambition', I turn to human aspirations and fulfilment. In part, the sense of meaning in our lives comes from the people who matter to us – from the smile of a child or the nod of appreciation from a colleague. But a part too comes from a sense of ourselves and what we can achieve. In Part III, I deal with self-respect – with the need to strive, to take up a challenge, to suffer, to be out on a fierce winter's morning, peering through the darkness towards the mountain top. The fierce need to strive and succeed is essential to our sense of ourselves . . . at the same time, it can run out of control in a modern, large-scale society. Corporations take on the needs of the individuals within them, striving for more and more billions even though further billions in themselves are pointless. The billions, I argue, are not truly wanted. What is wanted in our small, animal selves is the success.

Part IV is 'Sex'. In the twenty-first century, some of the most passionately debated social questions concern the roles and relations of men and women. On the one hand, we have the inspiring ideas of feminism and freedom of choice, with the ideal of equal opportunity for all to realise their dreams. On the other, we have the strong influence of the endocrine environment on brain development, and as any parent knows, often to their cost, it is not at all true that we shape our children entirely by the way we bring them up. With his usual penchant for abstraction, Spock may think that any recognition of difference between the sexes threatens equality of opportunity, 'equal is equal', but I shall argue that this is a mistake. Reality as always is more complex than Spock imagines, and to be fulfilled by their lives, I shall argue, men and women need both the Pan and the Spock of their dreams, their nature and their relations.

In Chapter 12, I deal with the difference between equal rights and identical people – the biology that makes male and female different animals, and the culture that gives both the chance to realise their dreams. Chapter 13 deals with sex and sexual attraction, with all their combination of conflicting, often unsettling desires and potential for deep satisfaction and meaning. In Chapter 14, I turn to a fascinating historical study of what happens when men and women really are raised in the same way and with the same expectations, what it tells us about the balance of our conflicting needs and desires, and what it means for our lives today. Once again, the story of twenty-first-century relations between the sexes is the familiar, sometimes screaming dialogue of Spock and Pan, with all its joys and all its challenges.

In Part V, 'Power', I turn to politics, and to Winston Churchill's remark, 'It has been said that democracy is the worst form of government – except all those other forms that have been tried.' It is easy to see why democracy is so necessary, but in the conflicting world of Spock and Pan, it is also easy to see why it can't possibly work that well. For many, 'democracy' and 'the will of the people' are sovereign in political life – more important, for example, than making things work out well for those people. But in the world of Spock and Pan, there is no real will of the people – for that matter, there is no real will of even a single person. In every head are conflicting voices calling out conflicting commands and, in democracy, the clever leader works out which voice to speak to. The leaders, meanwhile, might ideally be working for the good of the community, but all too obviously, this is often not the dominant element in their thinking. Out-competing the question of what is best are the burning needs of the individual to dominate, to win and, often, to acquire personal wealth. A better political system might down-weight these needs, but can such a system be found? For this, we need better knowledge in the hands of the electorate – but can that better knowledge ever be provided?

Know thyself

Different though we are, the story of Spock and Pan tells us that we are all in this together. A young girl sits in a garden in Nsukka, Nigeria; perhaps she is reading, perhaps she is dreaming; she will grow up to write luminous fiction and to be loved and hated for her advocacy of women's pride and freedom. As recounted by Eibl-Eibesfeldt in *Human Ethology*, in New Guinea, an Eipo man is sitting in the morning with others, warming himself in the sun, carving an arrow. He gets up, goes over to the women's group, picks up his child and brings it back to the men. For half an hour the child is the centre of attention, kissed, stroked, cuddled, entertained in high-pitched baby talk. In my own life, a young woman is hurrying in the kitchen; she has been out in the winter morning feeding calves, she has urged the boys outside to feed the pigs and made sandwiches for school, and now she wants to have a hot breakfast ready when my father comes in from milking. Our lives and our genes make each one of us unique, and as the girl from Nsukka, Chimamanda Ngozi Adichie, explains so beautifully in her online TED Talk, we each have many stories.[12] But in these stories there are always themes, and they are the themes of Spock and Pan, of human ideas built on animal instinct.

In this book, I want to show that the world of Spock and

Pan is exciting for its compelling, often rather simple answers to many of the great questions that puzzle humanity. But it is important too because these answers matter. In putting forward his views on human aggression, Lorenz reminded us of the Delphic oracle and the injunction 'Know thyself.' He argued that aggression – especially group aggression – can best be tamed by understanding, not by wishing it away. In *The Blank Slate*, Steven Pinker uses 'Know thyself' as the heading for an entire section. We would like our lives to be rich, fair, satisfying, safe, vibrant, meaningful, and for our best chance of approaching these ideals, we need understanding of our own nature. As we struggle to create a culture for our global, rapidly changing century, both Spock and Pan have their essential parts to play, and as we take sides in our strident culture wars, we might remember that, very often, these are really wars between different sides of ourselves.

Despite the human fear of our animal side, the story of Spock and Pan is not a story of Dr Jekyll and Mr Hyde. We need Pan to give us our values, our needs, our sense that life has meaning. We need Spock to generate his infinite world of ideas – sometimes great ideas, sometimes not, but with extraordinary flexibility and breadth. The two sides of our minds have quite different strengths, weaknesses, origins and operating principles. Our lives need both; together, I shall argue, they make us who we are.

PART I

Operating Principles

CHAPTER I

The animal lovers

Aiming for principles

In 1973, Konrad Lorenz, Niko Tinbergen and Karl von Frisch shared the Nobel Prize in Physiology or Medicine for their work in creating a systematic science of animal behaviour. Deeply thoughtful and ingenious experimental scientists, these people were also simple animal lovers. In the words of Niko Tinbergen in his Nobel lecture, they loved 'watching and wondering'.

Lorenz loved greylag geese, jackdaws, dogs, coral fish. In *On Aggression*, he describes a moment of theoretical insight from the frightened reaction of a greylag goose that he kept in his home. For those who have ever seen the state of the ground in a field of geese, this story has two remarkable aspects – the scientific deduction from the animal's moment of panic, and the idea that anybody, no matter how committed to his subject, would keep a greylag goose in his home.

Tinbergen experimented on grayling butterflies in the Dutch woodlands, using butterfly models on the end of rods to understand the stimulus features that attract males to females, and documenting the sudden startle response by which a butterfly, almost perfectly concealed when it sits with its wings folded up, flashes out the eyespots on its upper side at the approach of a predator.[1] Tinbergen's ideas on instinct came from stickleback fish, herring gulls and many other animals.

Von Frisch used observation hives with transparent windows, pots of syrup differentiated by visual cues, and individually marked worker bees, to analyse how a bee locates food sources, then in dances of astonishing complexity, communicates this information back to its fellows.[2]

All three spent long professional lives watching and wondering. Lorenz put it like this:

> Were it not for the unaccountable gloating pleasure some of us take in watching 'our' animals, not even a person endowed with the supernatural patience of a yogi could bring himself to stare at a fish, a bird or an ape with the unremitting perseverance which is necessary in order to perceive the governing principles prevailing in the behaviour of an animal.[3]

This sentence captures the two sides of ethology. On the one hand, there is the gloating pleasure of the animal lover. But beyond the gloating, there is the imagination: it is the governing principles that we are after. Nowadays there would be many ways to illustrate these principles, as the study of animal behaviour has become a large discipline. As I learned so much from Lorenz and Eibl-Eibesfeldt, however, I shall use many examples from *On Aggression* and *Love and Hate*. Where it is useful, I have added more from a comprehensive and lovely modern textbook, Rubenstein and Alcock's *Animal Behavior*.

The innate releasing mechanism

If we watch fish teeming over a coral reef, or a gaggle of geese beside a lake, or swallowtail butterflies swirling over a hillside, the first impression is one of chaos. It is a rapidly moving jumble, with little immediate impression of order. But with enough watching, order begins to emerge. The same fixed patterns of behaviour, made in response to the same external events, appear over and over again . . . in endless combinations, and with endless variation, but now seen as variations on the same, constant set of underlying themes.

A flying moth, approaching a light bulb on a summer's night, suddenly dives from view. A moment later, a bat flickers into the light; the moth detected its ultrasonic signal, and the sound released an immediate, organised defensive dive.[4] A finch chick stretches wide its mouth, exposing an elaborate, highly characteristic pattern of markings; seeing the right markings, the parent delivers food into the waiting mouth.[5] Over the coral, a damselfish is circling its territory; seeing an approaching member of the same species, sporting the same, elaborate pattern of colours and shapes, it launches an immediate ferocious attack.[6]

These constancies are everywhere, from insects to fish to birds to primates. Any nature documentary shows them in their endless variety, and a modern textbook on animal behaviour contains many hundreds of examples. Stimulus events of all imaginable kinds – predators, food, conspecifics, the weather, the time of year – call forth organised, often complex patterns of behaviour.

To explain these behavioural constancies, Lorenz proposed that the animal's nervous system must be filled with small processing units, like little pieces of computer code. Each piece of code detects a critical stimulus event and releases the corresponding pattern of behaviour. He called these bits of code 'innate releasing mechanisms' or IRMs, and the activating stimuli 'releasers'.[7] Throughout the animal kingdom, he proposed, IRMs drive the multiple instinctive behaviour patterns that an animal needs to survive and pass on its genes – pursuing food, avoiding predators, finding mates, caring for offspring, navigating home, migrating at a change of the seasons.

When Lorenz proposed the idea of the IRM in the 1930s, neurophysiology was in its infancy. Now, however, much is known about the detailed neural mechanisms of some IRMs. We know, for example, how the curious ears of a moth detect the approaching bat; we can trace the route by which this information is passed to the wing muscles, and follow the resulting collapse of synchronised flying and the dive to the ground.[8] In Chapter 4, I shall look at brain processes underlying the social behaviour patterns of mammals closer to ourselves but, for now, we need only the theoretical idea that, in the nervous system of each animal, these many IRMs must exist.

A beautiful feature of this idea is that complex behaviour is built up from simple parts. The first look into an open beehive will suggest a swirling chaos, but there is no chaos at all. The IRMs built into the bees' nervous systems are controlling precise roles – some are leaving the hive to search for food, some are instructing others where food has been found, some are tending the larvae, while the queen has mated and is producing eggs. As all these small fragments of behaviour are executed, the entire, successful activity of the hive is assembled.

The IRMs of the termite's nervous system dictate which grain of sand will be picked up, how it will be transported back to the mound, exactly where it will be deposited. As the multiple IRMs of millions of termites execute, an entire mound with its elaborate halls and chambers takes shape.[9]

On the simmering rocky shore of a Galápagos island,

36

Eibl-Eibesfeldt is watching a colony of nesting cormorants.[10] On each nest, an adult bird uses outstretched wings to guard its young from the sun. When a male returns to take over from the female, it must bring a small gift of new nesting material, which the female snatches and builds into the next. Only then is the male allowed to approach without being attacked. When he is ready to take over, he briefly holds the female's beak in his own, nibbles her neck and points his beak down into the nest. Now the female has been relieved and is free to depart. A concatenation of IRMs allows the cormorant colony to function and the individual pairs to successfully rear their young.

It is more than a loose analogy to describe the IRM as a fragment of computer code. Surely, what goes on in the beehive, the termite mound or the cormorant colony closely resembles the interactions of characters in a computer game, and is brought about in much the same way. In a computer game, too, individual objects and characters are programmed with their own properties and behaviour – a character by the roadside will give information if approached politely, but fight if offended; a door will remain closed if it is tapped at the bottom, but open when it is tapped to one side – and now there is the option of infinite variations on the story as these simple entities interact. In the case of animals, there is no external player, only the world's immense game playing itself, but inside the game, the broad principles of the protagonists are much the same.

Filling in the details

As indicated by the 'innate' in the innate releasing mechanism, Lorenz and the other early ethologists were concerned especially with instinctive behaviour. To test whether the IRM is encoded in an animal's genes, many biologists have reared animals under controlled conditions, where opportunities for learning can be varied or eliminated. Here is a typical example from Eibl-Eibesfeldt:

> The Central European red squirrel hides nuts in autumn as a provision for winter. In so doing it follows a uniform pattern: with the nut in its mouth it climbs down to the ground and looks about until it comes to the base of a tree trunk, scrapes out a hole with its forepaws, lays down the nut, presses it in firmly with its nose and finally scrapes back the loose earth over the nut with its forepaws. This behaviour pattern is not seen at all in baby squirrels, for they are born in nests, blind and naked. But I have several times reared squirrels in such a way that they had no examples to copy, and no opportunity to learn how to hide nuts by trial and error. Nevertheless these animals were able to carry out the species-specific hiding technique. When I gave nuts to the grown squirrels for the first time they began by eating them. When they had eaten enough they started to hide them. Each squirrel would run round the room with the nut in its mouth until it finally began scrabbling away in a corner; after that it laid down the nut, pressed it down with its nose and finally made the raking over and pressing down movements with its front paws – although it had not dug up any soil. This shows clearly that what we have here is the blind unfolding of a behavioural sequence which has been hereditarily programmed.[11]

Of course, however, there is also much learning in animal behaviour, likewise leading to order and constancy. A foraging wasp returns over and over again to the same nest; undoubtedly, genetically specified mechanisms have allowed it to learn its route, but it is the learning itself that allows repeated, orderly approaches to its own particular nest.[12]

As captured in the IRM concept, some 'instincts', such as the escape reaction of the flying moth, are likely specified in great detail in the genetic code, with little room for modification by the individual animal's experience. In other cases, the genetic code provides only a blueprint, with the details filled in by experience. This will become especially important in the next chapter, as we turn to human instincts and their modification by the surrounding culture, but in general, we should expect genetic codes to combine with individual experience to determine the final behavioural result.

A good example, reminiscent of human culture, comes from many years of experiments on the song of the Californian white-crowned sparrow.[13] These sparrows, of course, have their own species-typical song, but with recognisable variants along the length of the Pacific coast, like the dialects of human speech. If a sparrow is raised without hearing adults sing, its own song never forms properly, but is replaced by a sort of twittering. If it is exposed to the song of an adult sparrow during a critical period, between 10 and 50 days from hatching, then later, when it begins to sing at 150 to 200 days, it develops the full species-specific song – and in the dialect of the singer it was exposed to. If it is exposed to the song of a different species, again its own song is reduced to twittering. If it is exposed to two songs, that of its own species and that of another, again it learns its own, species-specific song, including its dialect. In this case, the genetic code evidently indicates broad features of the song that the bird prefers to learn, but the heard song of a specific tutor fills in the details.

In this sense the classic IRM, with a fixed action pattern released by a highly specific stimulus, may be seen as a common but special case. More generally, it is best to think of the genetic code as providing constraints on behaviour – sometimes very strong, sometimes only outlines, but explaining why it is that animals of the same species, even reared very differently, end up with such recognisable constancies in what they sense, learn and do.

With apologies to ethologist readers, in this book I shall use 'innate releasing mechanism' and the abbreviation I RM to refer to all these innate constraints, even those that serve only as blueprints to be filled in by learning. This is perhaps not stretching the concept too far; after all, we may well think that exposure to a suitable song releases the young bird's internal process of learning it. Anyway, this is what I shall do; hopefully it will not prove misleading.

Communication and ritualisation

One large class of I RMs serves the function of communication. In courting, competition, battle, care for the offspring, it can be important for signals to pass between one animal and another. A common element in the evolution of signalling is what ethologists call ritualisation – the term going back to Julian Huxley, and his studies of courtship in diving birds in the 1920s.[14] As related species evolve, an element of behaviour is coopted as a specific signal. In the process, the behaviour itself may be altered and exaggerated, assuming a specific fixed form; at the same time, the animal's body may become adorned with specific physical features rendering the signal more conspicuous and unambiguous. The ritualised movements, the bodily adornments and the response of the receiver must all co-evolve – serving the function of passing a clear message, of benefit for the sender to send and the receiver to receive.

In *On Aggression*, Lorenz used the 'inciting' of female ducks to trace the evolutionary shifts of ritualisation.[15] In many duck species, the females are as aggressive as the males, but smaller and weaker. Lorenz uses the common shelduck as a simple, non-ritualised case. In this species, the female may dash at a rival couple with outstretched, lowered neck, but then as she approaches, turn and dash back to the protecting male, where she may stand in front of him, to one side or behind, her head always pointing back at the rivals as she continues to threaten. Her display in turn incites the male, who also becomes enraged and joins the attack.

Lorenz then takes us through a variety of other duck species, in whom components of this behaviour have become stereo-typed and in some cases have taken on new meaning. If the female shelduck happens to be facing her mate as she continues to threaten, then inevitably her head is turned to face back at the rivals, and in the mallard, this turning has become fixed; even if she is standing in the wrong position, as her anger mounts, her head seems irresistibly pulled back to look behind her. In the mallard, furthermore, this ritualised inciting may be directed at a rival, but this is no longer its only function. Even when no rival is present, the same inciting movement can be used by the female as an invitation to form a pair, and if the drake is agree-able, he responds with his own characteristic signals, lifting his chin, turning away his head and uttering a specific call, 'rabrab'.

In *Love and Hate*, Eibl-Eibesfeldt tracks a similar progression in the courtship of the pheasant family.[16] To attract the attention of a hen, a domestic cock scratches the ground, pecks, and if he finds a crumb of food or even just a pebble, picks it up and drops it. In many animals, males use gifts of food to court females, and this is a straightforward if not always strictly honest example. But as ritualisation is traced through related species, the pecking at the ground becomes fixed in elaborate bowing and, at the same time, the bird's tail becomes exaggerated, enhancing the bow's effect. Eventually we reach the peacock, which simply spreads and shakes his enormous tail feathers, taking a few steps backwards. The hen still runs up and searches the ground for food, though now only the glamorously exaggerated bow has been offered.

The world of courtship is filled with ritualisation, from males of the cabbage white butterfly, displaying their nitrogen-rich yellow undersides in an elaborate courtship flight,[17] to male bowerbirds, building structures of flowers, pebbles and other brightly coloured objects to impress inspecting females.[18] So also is the world of aggression. If two animals, usually male, fight for territory, it is advantageous for both if the fight can be settled without bloodshed. An animal's size and weaponry determine its chance of success, and across the animal kingdom, elaborate displays have evolved that allow each combatant to display his strengths.[19] Male impalas guarding their harem approach potential rivals, snorting and shaking their heads to show off their horns. Fish battling for territory assume elaborate threat positions, waving brightly coloured fins. Some antlered fly species use face-to-face displays strongly reminiscent of threatening deer. Barking geckos indicate their size by the pitch of their bark. The battle is often ended when the aggressive display of one combatant changes to a ritualised display of submission – a Galápagos iguana falls flat on his belly, or a dog rolls over to present its unprotected underside, just as a puppy rolls over to be cleaned by its mother. In all these cases, the weaker combatant is able to withdraw before a real fight begins, showing the evolutionary merit of clear signals.

Disconnected fragments

In 1935, Lorenz published a paper entitled 'Der Kumpan in der Umwelt des Vogels'.[20] In my English edition, this is translated as 'Companions as factors in the bird's environment', but in his notes Lorenz remarks:

The German word 'Kumpan' denotes a concept somewhat different from that associated with the word 'companion'. The German word has a slightly derogatory meaning; a 'Kumpan' is not the companion of your soul but a fellow who shares your pleasure in hunting, drinking, frolicking, etc. (Saufkumpan, Jagdkumpan, etc.).[21]

With this idea of the *Kumpan* as a partner in just one specific activity, Lorenz was offering a bold proposal. In the mental world of a bird, with its many IRMs, there may often be no need for an integrated concept of a particular individual, such as a partner, a familiar enemy or a chick. Each correct behaviour is brought out simply by each particular releasing stimulus, and if the releasing stimulus is altered, no 'knowledge' of the individual can replace it. As each IRM is discharged, the bird sees just the *Kumpan* for this one particular act.

This is not always true. As we shall see in a later chapter, Lorenz also gave much attention to individual bonds in some species, especially his beloved greylag geese. Often, however, it is very true.

Perhaps this seems unsurprising and even obvious in simpler animals. All in a single moment, the female praying mantis copulates with the rear end of a male while turning to eat his front end. In this case, it seems right to think that the IRM of catching prey was inhibited just as long as the male's courtship continued, but as soon as copulation is in progress, he simply returns to the status of food. We would not presume that ants have any integrated conception of their sister ants; harmony is maintained simply because the particular scent of the colony inhibits attack.

But something very like this can also be true of birds. A female incubating her eggs must be in a highly aggressive state, primed to fight off any intruder; when the chicks hatch, how are they able to defuse the mother's aggression? In turkeys, the answer is the chick's call; if the chick is dumb it is immediately killed, while the unsuspecting mother will care for a stuffed polecat playing chick calls. The visual appearance of the chick is neither here nor there – if a deaf mother is shown a chick, 'she utters no call notes, but if the baby approaches within yards she raises her feathers defensively, hisses furiously and as soon as the chick is within reach of her beak, she pecks it as hard as she can'.[22]

In many bird species, when a parent returns with food to the nest, it is not 'the parent' or 'the chick' that elicit appropriate behaviour, but just simple features of each one. Famously, Tinbergen showed how begging herring gull chicks peck at the red dot at the end of the parent's bill, encouraging the parent to deliver regurgitated food, but will peck just as well at any red dot on a roughly beak-shaped stick.[23] Returning to its nest, the parent finch delivers food to a begging mouth with the right markings, but not to an equally hungry chick with no markings.

Competing IRMs

In an animal with many IRMs, triggered either by different releasers from the same conspecific, or simply by different features of the current environment, it becomes inevitable that there will be conflicts. Very likely, the turkey chick arouses both defensive aggression and the instinct to protect. Lorenz spoke of the parliament of the instincts – each instinct calling out to be satisfied, and with rules of dominance determining how this conflict plays out.[24] Julian Huxley called the animal 'a ship commanded by many captains'.[25]

A common case is the conflict arising when partners approach. On the one hand, an approaching animal is a potential threat. In many birds, for example, even other members of the same species will readily steal eggs or chicks from a rival's nest. On the other hand, approach can be necessary for joint activity, such as shared nest care or mating. In birds, commonly, specific ritualised behaviour patterns are used to ensure that an approaching partner is greeted rather than attacked. We have seen one example already, in Eibl-Eibesfeldt's description of the approaching male cormorant handing over a gift of nest material, and this use of gifts is widespread among birds. Eibl-Eibesfeldt also describes courting male cranes offering blades of straw, great crested grebes holding nesting material in their beaks, penguins and gannets using pebbles.[26] In many cases, clues to the competing aggression are seen in the way that the gift is taken, with a lunge closely resembling the lunge at an opponent. Even more tellingly, if the gift is omitted, the full attack occurs. Above we saw another good example in the mating praying mantis, with conflict between a male's status as partner or prey. To acquire the status of partner, he approaches the larger female with great caution, with an elaborate mating dance that, for her, increases the dominance of mating over feeding. Once mating is in progress, however, this dominance reverses and the male is consumed. In some spiders, the male suffers a similar transition, courting until he has mated but ending up as a meal.[27]

In all these cases, we must imagine that, in the animal's nervous system, the different fragments of code are competing to assume control of behaviour. Lorenz proposed this on strictly theoretical grounds but, in a later chapter, we shall see how exact neural mechanisms of conflict are now being unravelled.

Often, too, the competition can be seen quite concretely in the behaviour itself. As the competing control codes vary in strength, so does the animal's behaviour morph in a continuous space of combinations. Lorenz liked to illustrate this with a two-dimensional grid of drawings, showing these continuous changes in the strength of competing tendencies.[28]

45

A famous example is his depiction of fear and aggression in the face of a dog. The dog in his upper-left drawing is neither especially fearful nor especially aggressive. Its ears are pricked up, its mouth closed, its expression alert. Down the left-hand column of drawings, as fear increases, the ears go back. Across the top row, as aggression increases, the ears go further forward, the mouth opens and the face wrinkles in a snarl. Filling in the whole space, we reach the bottom right-hand corner, where the dog is both fiercely aggressive and fiercely afraid. The ears are back, the face is snarling, the dog is in the last extremity of desperate defence.

Most of us are familiar with dogs, and the expressions that Lorenz depicts are all quite recognisable. In the work of Lorenz and his research group, however, much the same methods were used to understand new, unfamiliar cases – the exact movements of greylag geese that showed the balance of fight and flight or, in cichlid fish, the competing tendencies to fight, flee or mate. The more aggressive goose stretches its neck forward and up, the more submissive goose kinks its neck down so the head approaches the ground. In the cichlids, the tendency to fight is indicated by body orientation towards the other fish; the tendency to flee is indicated by body orientation away; the tendency to mate is indicated by orientation towards the ground, linked to the need to dig a nest hollow and spawn. Tinbergen called this technique 'motivation analysis'. Using a thorough motivation analysis, ethologists found they could predict the outcome of an interaction – for example, the meeting of two rival ganders – based on the positions that each one adopted.

In the parliament of the instincts, it is not just the major drives that compete – fight versus flight, hunger versus sexuality. Lorenz proposed that competition runs all the way down, with each individual movement pattern competing to take control of the animal's behaviour. Accordingly, the 'releaser' does not just release the matching behaviour; rather, it calls for that behaviour, with the final result always determined by what other competitors, activated by their own releasers, are simultaneously in play.

Competition as a broader principle

Beyond ethology and the parliament of instincts, conflict and competition are central ideas throughout psychology. Every moment of our waking lives is filled with decisions, from the small decision whether to push back the sheets or have another moment in bed, to the large decision of whether to have children or devote more time to career. To frame thinking about these problems, it is often useful to imagine that each alternative has a competitive strength or weight. The greater the weight of one alternative, the more likely it is to be selected – though just as we see in external conflicts, there is always some variability in outcome, and the strongest competitor does not win every time.

Thinking this way, we can begin to analyse the conditions that make each alternative stronger or weaker. Often these are called weighting functions. One alternative may be stronger because it fits in with a current plan. Another may be stronger because it is highly familiar. In his 1890 book *Principles of Psychology*, one of the fathers of modern psychology, William James, recounts once going up to his bedroom intending to change clothes for dinner; taken over by habit, he found he had undressed and got into bed.

A simple laboratory example, first described by the psychologist J. Ridley Stroop in 1935, has since been used in countless experiments on the nature of conflicts.[29] The person is shown strings of symbols, each string in a different colour, and asked to name the colours. This is easy when the symbols are rows of Xs, but suddenly becomes very difficult when they spell the wrong colour words, such as BLUE written in red ink. The person hesitates, stumbles, occasionally even reads the word rather than naming the colour. The name of the colour gains weight because it fits in with the current plan; the written word has strong weight because reading is so familiar; just like fight and flight in the dog or goose, the two tendencies compete to take control of behaviour.

In everyday behaviour, the weight of an alternative fluctuates up or down as surrounding conditions change. As an opponent stretches up his neck and appears larger, the weight of attack decreases and the weight of flight increases; as the frightened animal turns and flees, the weight of flight sinks again with increasing distance from the terrifying stimulus, and the animal may turn and come back for more. From relatively simple weighting functions, complex, constantly evolving behaviour can emerge.

In Chapter 4, we shall consider modern experiments on how this plays out in the brain. As Lorenz imagined, we see neural generators for competing lines of behaviour, each varying up or down in strength as eliciting conditions change, and each tending to inhibit the other.

Weighting functions provide a useful way to think about geese, fish and the parliament of instincts. They will become even more useful when we turn to the human, and the complex world of Spock and Pan. Competition is not just limited to explicit decisions on what to do next; just as strongly, it exists in the world of ideas, bringing the same rich mixture of opposing, evolving, often frankly inconsistent thoughts.

The need for discharge

The last great idea of Lorenz was that each behavioural tendency – each competitor in the parliament of instincts – *needs* to be discharged. With the passage of time without satisfaction, this need increases, along with the instinct's competitive strength. Eventually, the behaviour may appear even with no suitable releasing situation. Lorenz illustrated his ideas with a hydraulic analogy, with water gradually filling a tank and the releaser allowing the increasing pressure to be discharged. Nowadays we know that this is roughly right, though rather than increasing water pressure, what increases is the excitability of sets of neurons in the brain.

In some cases, of course, we are all too familiar with this spontaneous need for discharge – think just of a period away from one's sexual partner. A predecessor of Lorenz, Wallace Craig, traced the equivalent in the ringed dove, removing the female from the male for increasing lengths of time:

> A few days after the disappearance of the female of his own species, the male was ready to court a white dove which he had previously ignored. A few days later he was bowing and cooing to a stuffed pigeon, later still to a rolled-up cloth, and finally after weeks of solitary confinement, he directed the courtship towards the empty corner of his box-cage.[30]

Lorenz proposed this spontaneous build-up as a general rule, applying from major instincts like the sexual drive to individual, specific IRMs.

Most controversially, Lorenz proposed a growing need to discharge aggression. Unsettling though this need may be, it is clearly seen in many animals. Eibl-Eibesfeldt discusses the case of the male cichlid *Etroplus maculatus*.[31] If a male has been kept alone in an aquarium, it is impossible to mate him with a female; when she is put into the tank, he fiercely attacks. To resolve the situation, he must first be given another male to fight; having attacked this male, he is now ready to court the female. Now imagine that two pairs are put into the same tank, with a pane of glass preventing their coming together. The males can threaten one another through the glass and live peacefully with the females. But if one pair is removed, the aggressions of the remaining male build up; eventually he kills his partner.

Similar considerations apply to catching prey and feeding. Here is another example from Lorenz:

> A hand-reared starling that I owned many years ago had never in its life caught flies nor seen any other bird do so . . . One day I saw him sitting on the head of a bronze statue in my parents' Viennese flat, and behaving most remarkably. With his head on one side, he seemed to be examining the white ceiling, then his head and eye movements gave unmistakable signs that he was following moving objects. Finally he flew off the statue and up to the ceiling, snapped at something invisible to me, returned to his post and performed the prey-killing movements peculiar to all insect-eating birds. Then he swallowed, shook himself, as many birds do at the moment of inner relaxation, and settled down quietly. Dozens of times I climbed on a chair, and even carried a step-ladder into the room . . . not even the tiniest insect was there.[32]

The passage of time is one factor increasing the need of an IRM to discharge; equally, it can be aroused by some releasing event but also inhibited, leaving the need for discharge elsewhere. Redirected aggression is a well-known example; an animal is attacked by a higher-ranking member of his group, to whom he is unable to respond; in response he seeks out a lower-ranking alternative and attacks him instead. Changes in the strength of a behavioural tendency can also accompany a change in the seasons. Migratory birds are increasingly restless as the time for migration approaches. In captivity, they show an increasing tendency to take off in the direction they would take if they were free.[33]

As a prisoner of war, Lorenz had experienced at first hand the increase of aggression in a small band of isolated men forced to live apart from strangers. Rather than attacking his companions, he thought the well-informed ethologist should find an outlet by creeping out of the tent and smashing some object with a resounding crash.[34]

Pan's rules

Here, then, we have Pan. He consists of a large body of IRMs, each bringing forth particular, sometimes complex patterns of behaviour in response to particular, sometimes simple, environmental events. He is not coherent – often his behaviour is driven by just one small aspect of another animal or a situation, with no fixed conception of a wider whole. He is conflicted – the same partner can bring out quite opposite behavioural tendencies, depending on the particular releasing stimuli in evidence and the particular surrounding circumstances. He is inconsistent – as time passes, each instinct varies continuously in strength, and what is dormant at one time may become insistently dominant at another. In short, he is a jumble of many fragments, rather like the animal's life itself.

When we read this description, we are surely reminded of aspects of ourselves. In the next chapter, we shall look in detail into the case of *Homo sapiens*.

CHAPTER 2

Obviously

Us too

It perhaps seems far-fetched to imagine that the principles of animal behaviour, from insects to birds to chimpanzees, would suddenly cease to apply to *Homo sapiens*. Yet throughout the twentieth century, this was exactly the received wisdom in the social sciences. In *The Blank Slate*, with quote after quote and example after example, Steven Pinker documents the belief that human nature is the product of learning and culture, not evolution. To believe otherwise, indeed, has often been seen at best as naive, at worst reprehensible.

How likely is it that one species would have no innate releasing mechanisms – along the lines of the IRMs we looked at in Chapter 1 – no conflicting instincts, no significant innate constraints on their behaviour? Innate behaviour patterns arise through stability in an animal's evolutionary history – conditions and requirements that remain stable for sufficient time that useful behaviour patterns have become hard-coded in the nervous system. If *Homo sapiens* really evolved in a world with no fixed constraints, only with conditions that changed from generation to generation and from culture to culture, then perhaps it would be conceivable that the nervous system was left open, to be entirely formed by individual experience. From this thought it is perhaps natural to move to the realisation that, indeed, *Homo sapiens* does live in a world of extraordinary variability, and is a creature with extraordinary learning and adaptability. Of course, a modern city dweller faces requirements very different from requirements on, perhaps, the !Kung Bushmen of the Kalahari or the Yanomami of the Venezuelan rainforest. To manage their lives, these three groups of people need banks of knowledge that at first sight seem almost entirely different.

But is *everything* different? We academics are often mocked for our ability to defy common sense, and common sense tells us that, if we meet a !Kung Bushman or a Yanomami mother, this will not be like meeting an alien from another world. There will be smiles, possibly threats, gifts, hugs, uncertain laughter, interest in one another's children . . . and if common sense does not tell us this, then Eibl-Eibesfeldt, with the material to follow in this chapter, soon will. I think we academics are prone to miss complex reality exactly because we rely so heavily on Spock – on the reasoning machine and its abstractions. Spock finds a useful idea – perhaps the idea of immense human adaptability and variation – and then he begins to apply it. First, perhaps, he applies it where it works but, soon, the abstract idea is applied too generally. The idea is simple and powerful, but reality can be complex and muddy, and the more muddy it is, the more likely it is that Spock's abstract ideas will go wrong.

The worlds of the modern city dweller and the tribal hunter-gatherer are different in many ways but, across cultures, there is also remarkable consistency in social signals, social interactions and social structure. Already in *The Expression of the Emotions in Man and Animals* (1872), Darwin pointed to the similarity of facial expressions across cultures. Almost a century later,

Eibl-Eibesfeldt made an extensive attempt to analyse behavioural similarities and differences across many cultures, including those barely touched by contact with external civilisation.

In long field trips, he gathered data among the Yanomami of the South American rainforest, the Eipo of New Guinea, the Bushmen of the Kalahari, the Himba of southwest Africa, Trobriand Islanders, Balinese. Aware that a person's behaviour may change when they know they are observed, he made extensive use of a camera that photographed round a corner, appearing to point one way but actually recording in another. Then the resulting films were analysed in painstaking detail, documenting precise movements with their exact timing as men, women and children navigated through the complex social interactions of their daily lives.

In a monumental summary, *Human Ethology* (1989), Eibl-Eibesfeldt presents the results of this lifetime's work. The book is filled with sequences of film stills showing Yanomami children squabbling over a stick, Eipo women calming a crying child, a G/wi Bushman baby flirting with a stranger, a modern Frenchwoman greeting an acquaintance with a sudden upward flick of the eyebrows. The film sequences are completely persuasive of one thing – these are not aliens. Looking at the film you know exactly what is going on. If you were asked to babysit that child, you would know how to do it. For at least several million years, humans and their predecessors have been an intensely social species, dealing with allies and enemies, social superiors and inferiors, mating partners and offspring. Inevitably, stable requirements on social relations have created stable, heavily instinctive social minds.

Once again, we must remember to check Spock in his tendency to run away with an idea. In the case of birdsong, we considered how genes can provide just an outline blueprint of required behaviour, with the details then filled in by learning. In the case of the *Homo sapiens*, with our truly immense learning ability, this dominates the picture. Very often, we see a recognisable common theme across the behaviour of different cultures, but with many, sometimes extreme variations. The ethologists of course recognised the close relationship between biological and cultural ritualisation. In biological ritualisation, as we saw in Chapter 1, behaviour becomes increasingly stereotyped through evolution, allowing an unambiguous signalling function. In cultural ritualisation, instinctive patterns of behaviour are modified and then copied across the generations, producing a rich range of variants. Though fixed within a culture and varying from one culture to the next, these variants remain recognisably the same in their broad form and function.

Often, too, this similarity is not only visible between human cultures. As we saw so often in Chapter 1, similarity also extends *across* species, as similar evolutionary demands lead to similar behavioural solutions. Though human social life is perhaps more complex than that of any other species, and though each species has its own unique adaptations, sometimes our displays of defence, appeasement or affiliation can be strikingly like those of other animals.

Some examples: Babies, shivers, the kiss

A compelling first example of human IRMs is the baby schema. As we saw in the last chapter, for many animals, the caring response of the parent is released by a specific sign stimulus from the young – the cheeping of a turkey chick or the brightly coloured throat of a begging baby finch. Early in his writings, Lorenz pointed out that specific features of human infants release our own caring response.[1] These features include short, chubby legs, a head that is large relative to the torso, a bulging forehead and large eyes. We see this and we want to cuddle and protect it.

Often, IRMs can be triggered especially strongly by artificial, supernormal stimuli. A good example comes from Tinbergen's studies of herring gulls, oystercatchers and other birds.[2] If an egg has rolled out of the nest, the mother bird uses its chin to roll it back, and to a large extent, the bigger the egg the better. If a mother bird is given the choice between a normal egg and a beachball, it is the beachball she chooses to roll back, struggling with the enormous object against her breast. The same happens with the human infant schema. If people are shown drawings of baby heads, it is heads with artificially increased baby features – larger heads, more bulging foreheads – that are preferred. They look cuter – and as ethologists were quick to point out, the appeal of these baby features is easily seen in dolls and cartoon characters. Images of Mickey Mouse over 50 years show how he slowly evolved, with eyes growing to an absurd size, and thin mouse legs evolving into the short, chubby legs of a human baby.[3]

Another example illustrates several core features of human IRMs – communicative function, evolutionary history and cultural variation. Here, we turn from care to threat. We all know the 'shiver' that runs through us in the face of a challenge to be met – but where does it come from? The shiver is a piloerection response, tending to make the hair on our shoulders and outer arms stand on end.[4] It is often accompanied by raising the shoulders, rotating the arms inward and standing more erect. In the hairy chimpanzee, this is all easily seen as a series of tricks making the body look larger, used when the animal defends its band. As Lorenz says, it is 'a bluff'. In us, the hair has largely gone, but the old ape piloerection response remains in the form of a peculiar, sometimes almost spiritual shiver of anticipation.

The hair has largely gone, but the attempt to make the shoulders larger has remained in countless cultural variants, many associated with male threat. Eibl-Eibesfeldt illustrates this with the epaulettes of a Russian tsar, the costume of a severe Japanese kabuki actor, and a Yanomami Indian, with huge clusters of spiky feathers attached to each shoulder. When we see enlarged shoulders, we know what it means. It means 'Don't mess with me.'

A third example of human IRMs is kissing. Eibl-Eibesfeldt proposed that human kissing – between adults and children, or between one adult and another – has been ritualised from kiss-feeding. *Human Ethology* has images of kiss-feeding across cultures – !Ko Bushmen, Yanomami, Blit, G/wi and many more. In kiss-feeding, the mother passes chewed food to the child but, often, the same movement is used simply to pet the child. 'In this case the lips are placed on the child's lips, and the tongue is pushed briefly between the receptively opened lips of the child, without transmitting anything more than saliva.'[5] Children also kiss-feed one another, or babies kiss-feed their carers. We see a Yanomami girl kiss-feeding her young sibling, and a Himba baby kiss-feeding her grandmother.

Kiss-feeding is, of course, strongly reminiscent of parental care in many species, such as many birds feeding the young from their beaks. In the human case, the kiss is not simply a means for transferring food; it has been ritualised into an action taken for its own sake, strengthening the bond between partners. A similar social role is seen in many carnivores, who use gestures of nuzzling mouths, begging for food and rubbing noses in greeting and appeasement.[6]

From kiss-feeding it is a small step to the kissing we know

among adults, in both sexual activity and (usually in abated form) in all kinds of friendly greetings. Like kiss-feeding, kissing as part of sexual activity is also widespread across cultures, sometimes accompanied by actual kiss-feeding. In *Human Ethology*, Eibl-Eibesfeldt gives examples of sexual kissing and kiss-feeding from ancient Japan, Peru, Papua/New Guinea. He points out that apes also kiss-feed their young, and chimpanzees also use brief mouth contact in greeting.[7]

In the 1990 film *Pretty Woman*, the rather implausible sex worker played by Julia Roberts says that she will have sex with her client, but no kissing . . . and without thinking about it, we kind of know what she means. While intercourse can just be a no-frills act of physical pleasure, the deep swoon of a protracted kiss is something else – an act of bonding and intimacy. Perhaps this reflects its deep roots in sharing and care, initially between parents and infants, then ritualised into bonding between adult mating partners.

Facial expressions

In the human being, it is the face that is most obviously used for stereotyped social communication. The human face contains more than 40 individual muscles (the number depending on which are included and how finely they are separated), allowing us to alter the face into many different configurations. These configurations are instantly recognisable for the emotional states they communicate.

The idea of 'basic', easily recognised emotions has been developed in great detail by the psychologist Paul Ekman.[8] As Darwin already recognised, basic emotions are captured in much the same expressions across cultures, and sometimes across species. A threatening mandrill turns down the outer corners of its mouth in a snarl, showing its dangerous canine teeth. Threatening humans do the same: though the large canines are no longer there, the corners of the mouth are still pulled strongly down – illustrated in Eibl-Eibesfeldt's *Love and Hate* by a young girl in a flaming temper, and by a Japanese actor portraying extreme rage.[9] At the same time, the eyes stare, the cheeks are furrowed up – the whole expression immediately communicates fury.

In disgust, the nose is wrinkled, the forehead furrowed, the eyes almost closed in rejection. Eibl-Eibesfeldt shows an Eipo man rejecting a clove of garlic – forehead wrinkled, eyes half-closed, nose pulled up, head turned rejectingly away – and similar rejecting movements are made by small children given sour food.[10] The facial expression of disgust is similar whether the disgust is prompted by sour food or moral scorn, and its form strongly suggests ritualisation from the basic movements of rejecting unpalatable food.

Other basic emotional expressions in the Ekman scheme include fear with its staring eyes and tightly drawn mouth, surprise with parted mouth and eyebrows pulled up, as if pulling open the eyes to drink in the scene, sadness with the face crumpled and happiness with a beaming smile. This beaming smile is often used by infants as they playfully strike another child or adult, or make some other invitation to play, and resembles the 'play-face' of tussling chimpanzees.[11]

As we would expect for communicative signals, ritualised through evolution, these expressions are stark, stereotyped and easily recognised whether they are made by a close friend or a child from the Amazon rainforest. To establish similarity across cultures, Darwin drew up a candidate list of stereotyped emotional expressions.[12] Items included, 'Is astonishment expressed by the eyes and mouth being opened wide, and by the eyebrows being raised?' 'When a man is indignant or defiant does he frown, hold his body and head erect, square his shoulders and clench his fists?' He circulated his list among missionaries and others closely acquainted with 'several of the most distinct and savage races of man'. The replies persuaded him that 'the same state of mind is expressed throughout the world with remarkable uniformity'.

Expressions do not need to be imitated or learned. Children born deaf and blind will smile and cry,[13] and in a rage, bare their teeth, clench their fists and stamp their feet.[14] Darwin, too, pointed out how children born blind will blush from shame.[15]

Recalling the principles of releasing stimuli in many species, furthermore, these stereotyped social signals release a similar response in the receiver, whether or not they are actually made by a human face. The camel appears arrogant because it lifts its nose in an apparent sign of rejection. The eagle appears noble and aloof with its shaded eyes and narrow, downturned mouth. Releasing the sense of respect, across history the eagle has been used as a symbol of nobility on flags, statues and aristocratic insignia. With these reactions, we are very much like the oyster-catcher that attempts to roll even a beachball back into its nest, or the herring gull chick that is constructed to peck at the red dot on its parent's bill but, just as happily, will peck at a red dot on a yellow stick.

. . . and much more

Facial expressions play a crucial role in communication, and we should be hard pressed to navigate our social lives without the instinctive ability to produce these signals, coupled with the instinctive ability to recognise them. The complex social life of human beings, however, calls for much more than this, and indeed, much more is suggested by Eibl-Eibesfeldt's photographs in *Human Ethology*. The story told by these photographs is not one of a few simple, perhaps undesirable remnants left from an animal past. It is one of a complex system of social behaviour patterns, central to every aspect of our daily lives together. Across cultures, we see all kinds of stereotyped social interactions, affectionate and aggressive, blatant and subtle, with initiators and responders, from innocent babies to sophisticated adults.

We see comforting: a Yanomami mother wheedles her infant to stop crying; an Eipo woman hugs a crying child and points to the camera as a distraction; a sobbing American soldier buries his face in the chest of a comrade, who sweeps him into a comforting embrace; a frightened young Sono child clings to an older one, just as a frightened rhesus monkey clings to its mother.

We see stereotyped greeting: few of us are aware that, when we pass a friend in the street, we flash our eyebrows upwards for about a third of a second, but the same movement is seen in Europeans, Yanomami, !Kung, Huli, Balinese and Eipo, and when we see it, we do not think that the eyebrows have been flashed, we think only that we have received a greeting. A slow raising of the eyebrows, in contrast, indicates disapproval and rejection, in this case captured in the phrase 'raising eyebrows'.

We see a Yanomami woman flirting, with a sultry stare and flick of her tongue. We see clearly recognisable challenging and mocking from the Yanomami to !Ko Bushmen, sometimes bent over with bottom raised towards the victim, sometimes with hands challengingly thrust on to hips, sometimes with tongue thrust slowly out. We see a Yanomami boy proudly strutting and striking himself on the chest, an Australian aborigine admonishing with raised finger, a Himba boy and a Balinese mother delivering the slow, wide-eyed threat stare.

Even toddlers understand very well how to use these tricks

to manage their social lives. Two Trobriand girls play with a ball. After the ball has been repeatedly exchanged, one keeps it; she watches carefully how her friend reacts, provocatively offers it then pulls it back; now the other has it and teases her back, and there is pouting and turning away. Two Eipo girls are playing with a blade of grass; one playfully takes it away and the other disgustedly turns her back. An Eipo infant on its mother's shoulders opens out its arms to invite a hug from a young passerby, its face beaming with invitation. A Yanomami mother has ignored her son's attempt to draw her attention; when she finally turns to him, he immediately turns away in protest. Yanomami, Eipo, Bushman and Trobriand mothers entertain their babies and toddlers by making faces, with eyebrow flashes, mock surprise, playful grimaces.

It is all subtle, sophisticated, elaborate and finely balanced, and we understand it in a way that we could never understand the social signals of an insect, bird or other mammal. Despite the innumerable differences between one culture and another, these photographs by Eibl-Eibesfeldt tell us that, not far underneath the surface, we are all one.

Language and theory of mind

It is striking to see how much our social signals and behaviour can share with the signals and behaviour of our primate relatives. Human beings, however, also have social relations quite unlike those of any other animal, and these too have their strong IRMs.

Most obviously, we have language. Academics may debate whether other species have language, but this is really not the point – whatever language another animal may have, it is absolutely nothing like ours. Academics also debate exactly how much of the structure of language is instinctive, looking for common or universal elements across all human languages. This is fascinating, but what is beyond doubt is that all humans are born to learn and use language for detailed, complex, infinitely variable communication of ideas. This is something that allows humans to share information in a way that no other animal can, matching our own, species-specific, information-rich ecological niche. A human born without language would be massively dis-advantaged in living a human life; correspondingly, the need to learn and use language has been built into our genes.

Less obvious, but at least as important, is what social psy-chologists call 'theory of mind'. It is so natural to us that we are unaware of it, but each of us has a jaw-dropping ability to understand something elaborate, infinitely variable and largely hidden from our senses – what another person is thinking, wanting or planning. We know that our wife wants a cup of tea; we can offer to make one, with all the social advantages that brings; we can wait until she snaps and offers to be the one to put the kettle on; if we wish to let her know that she offended us earlier in the day, we can just make our own. We suspect that our neighbour has noticed our tree shading his vegetable garden; we can ring his doorbell and ask if he would like us to cut it down.

No other primate does this in remotely the same way, so no other primate can collaborate, plan or navigate their social relations in anything like our way. Like language, this ability to understand the minds of others, and to make use of this informa-tion in planning our daily lives, is special to us. As the psychologist Michael Tomasello puts it, 'It is almost unimaginable that two chimpanzees might spontaneously do something as simple as carry something together or help each other make a tool.'[16]

Much of Tomasello's work documents how early in life this 'theory of mind' develops, how complicated the theory soon becomes, and how special this is in human infants. A 14-month infant observes an adult bending over to click a light switch with their head. If the adult's hands were busy with something else – perhaps huddling in a blanket – then the infant, when it comes to their turn, just uses their hand; but if the adult's hands were free all along, the infant is more likely to follow suit and also use their head. The infant understands that no adult will butt a light switch without a good reason! Adults easily see social interactions between dots on a screen, for example if one appears to chase the other, and so do 12-month-old babies; they can tell the difference between a dot that helps or hinders another trying to climb up a slope.

How special this all can be is made clear by careful, often very surprising comparisons between apes and human toddlers. In one study, Esther Herrmann and colleagues compared toddlers, aged two-and-a-half, with older chimpanzees (average age ten) and orangutans (average age six).[17] Some tasks concerned the physical world – remembering where a reward had been hidden, discriminating quantities, understanding causes, using a stick as a tool. Others involved the mental world – learning by observation, following gaze to something important, understanding what an actor was trying to do. For the physical tasks, the apes were just as good as the human children, but for the social tasks, even at age two-and-a-half, the humans were already much, much better.

Just as we are born to use and learn language, so are we born to understand the minds of others. Of course, the way we learn language depends on the particular social world we grow up in and, very likely, the same applies to theory of mind. Experience is needed to fill in our language and our understanding of others, but this experience reaches a mind programmed to receive it.

These essential aspects of ourselves may not be the highly stereotyped 'instincts' of the fixed action pattern, but still, they build on genetically prepared origins, shaped through human evolution and making human life possible.

Conflicts

Like any other animal, the human must deal with conflicting IRMs. There is not a single, coherent person; instead, there can be opposite inclinations and thoughts, each with a weight that varies up and down with time and circumstance. We well know that we may wish to share a delicious snack but also to keep it for ourselves, or to stop and chat or to rush home for lunch, and as in the dogs, geese and cichlids of Lorenz, some of the most striking examples of such conflicts come in the management of approach, aggression and appeasement.

In the photographs of *Human Ethology*, the conflict of approach and retreat is seen vividly in young children, especially when they are confronted with a stranger. A G/wi Bushman baby sees that it is being watched. It hides its mouth with its hand, peeps and smiles, turns quickly away, rapidly alternating between contact and withdrawal. An Eipo baby looks, smiles, hides its face in its mother, looks back, then comforts itself at the breast. A Balinese girl smiles shyly, looks away, looks back again and beams. Two Tboli girls receive a compliment; they smile, one hides her mouth in her hands, they look away, look back. Of course, this is not restricted to children. An Eipo woman receives a greeting; she alternates back and forth between beaming, looking away, hiding her mouth behind her hand.

In humans, as in many animals, the teeth are potential

weapons, and as we saw earlier, a display of teeth is used in facial signals of anger and threat. This human action of hiding the mouth reminds us strongly of appeasements in many animals, allowing approach by hiding stimuli that release aggression. Eibl-Eibesfeldt uses the example of the blackheaded gull:

> Blackheaded gulls have great difficulty establishing friendly contact in the first stages of pair formation. Both partners bear a black facial mask, which is an aggression-releasing signal. If the female wants to approach the male, she may not look directly at his face. She approaches him in a stooped posture with intensive begging movements, which inhibit aggressive responses. The head is repeatedly turned away in a ceremony called 'head flagging' . . . Only when the gulls become individually acquainted . . . can (they) approach and look at each other directly without an appeasement ceremony.[18]

In many animals, from lizards to puppies, an appeasement gesture involves falling to the floor, or even rolling on to the back. Aggression is inhibited by a ritualised movement of declining conflict and even, potentially, inviting attack. Throughout history and across cultures humans have approached dignitaries and rulers by bowing, touching the forehead to the ground or throwing themselves to the floor.[19] The exact form varies over cultures, but the common theme is unmissable. Eibl-Eibesfeldt, indeed, suggests that nodding the head in agreement is a further ritualised form of this same movement, now as it were 'submitting' to the speaker's thought, and lists the many cultures in which he has observed nodding during greeting. In chimpanzees also, a subordinate may bow their head in greeting to a higher-ranking animal, and the gesture is returned by a hand laid on the subordinate's head. Similar to this return gesture, in humans religious and other leaders practise the 'laying on of hands'.

Eibl-Eibesfeldt suggests that, when groups come together, elaborate cultural differences are built on a common underlying theme of conflict management.[20] On the one hand is the need to impress, to show that our group must be taken seriously. One village of the warlike Yanomami is visiting another to cement the bond between them and negotiate mutual benefits. As part of the arrival ceremony, the visiting warriors dance with their weapons, often aimed directly towards the hosts; behind the men, children dance, waving palm fronds as a sign of peace. In a Balinese greeting dance, boys threaten with spears, while girls offer flowers.

Once the theme is seen, it is all absurdly reminiscent of our own, industrialised greetings between countries: a military display, a salute and a child presenting a bouquet. Or a firm hand-shake, a slap on the back, a compliment and a smile. A gift can promote the bond, but a gift can also be a threatening display of dominance. To defuse this we do a favour and say, 'It's nothing.' A Yanomami man gives a beautiful dog and says, 'Take this ugly creature.'[21] Every social interaction is a matter of fine balance, and without thought we balance all day every day, using just the same methods as the Yanomami, the Balinese and meeting heads of state.

Fluctuations

Innate releasing mechanisms compete for control, making a person assertive at one moment, submissive the next. As Lorenz proposed, we are well aware how the strengths or competitive weights of competing behaviour patterns vary with time and circumstance. Our irritability fluctuates with the weather, the amount of time since we were fed or the accumulating annoyances of a frustrating day. The increase in our need for sex with the passage of time resembles that of many other animals, who eventually, as we saw in Chapter 1, will be led to court quite unsuitable partners. All of this fits well with a Pan – the innate, instinctive side of ourselves – who is not consistent or coherent, but has a set of many competing thoughts and tendencies, some directly contrary to others.

In the case of sex we understand all too well the spontaneous need for discharge, and the way it builds up over time, but a little thought shows that the same applies to many different social instincts. When the children grow up and leave home, the space left in our hearts is filled with astounding ease by a new cat; though this largely non-social animal is quite incapable of understanding or responding to the elaborate care showered upon it, we speak to it in baby talk, speculate on its thoughts, tell it what is happening and are half convinced its gaze contains trusting affection.

The same applies to our simple need for company. After the first wave of the COVID-19 pandemic, many of us found ourselves returning to the office after perhaps a year of working almost entirely at home. For me, that year at home was perfectly acceptable – I had a pleasant space to live and work, a garden to look out on, an entertaining wife, a job that could perfectly well be conducted from home. Accordingly there was no strong urge to return to the office but, when I did return, one thing rather astonishing was that I had become quite spectacularly chatty. Generally speaking, if I am at work I work, but for at least a few days, I was pretty much ready to talk to anybody. Rats are just the same. Playing rats wrestle, sniff, groom, box and pin one another to the ground, and they play especially hard if they are put back together after a few days of isolation. They have been on their own, and now they're ready to party.[22]

The principles apply to us too

Obviously, the principles of instinctive animal behaviour apply just as firmly to ourselves as they do to insects, birds and chimpanzees. From the simple fixed action patterns and releasers of our facial expressions, to the elaborate means by which we navigate an interaction with an angry child or the approach to a powerful sovereign, we resemble any other animal.

Certainly, the way that instinct plays out in human life is complicated. Partly, this is because our social life itself is complicated, with the most subtle blending of roles, relationships, affections, responsibilities, all with new layers of depth added by our use of language and our ability to imagine the detailed content of another's mind. Partly, it is because, in our case, instinct is constantly remodelled by culture, with evident common themes, but seen through countless cultural variations.

Underneath all this complexity, however, there is something much the same. Here is a nice quote from William Hamilton, one of the deep thinkers on the evolution of social behaviour:

> Every schoolchild, perhaps as part of religious training,
> ought to sit watching a *Polistes* wasp nest for just an hour . . .
> I think few will be unaffected by what they see. It is a
> world human in its seeming motivations and activities far
> beyond all that seems reasonable to expect from an insect:
> constructive activity, duty, rebellion, mother care, violence,
> cheating, cowardice, unity in the face of a threat – all these
> are there.[23]

All this is built into the wasp, and all of it is built into us. This side of ourselves may have evolved in a world very different from the one we inhabit today; it may be limited, inconsistent, irrational, short-sighted; but with all of these failings, it is one indispensable part of who we are.

The sense of meaning

And, it is not just that we have our animal side, our Pan. This side, I believe, is essential to all that we want and need from our lives. Without this side, our lives would be meaningless.

'Meaning' was the core idea of the Austrian psychiatrist Viktor Frankl, founder of what is sometimes called the third school of Viennese psychiatry (the first two being led by Sigmund Freud and Alfred Adler). As a Jew, Frankl was interned when the Nazis annexed Austria, and spent the war in the increasing horrors of the concentration camps. He describes how, arrested with his family in 1942, he was clutching the precious manuscript in which he had explained his thinking, still hoping that it might be possible somehow to publish it. He could have no idea how life would be changed, first in Theresienstadt and finally Auschwitz, where his mother and brother were killed in the gas chambers. Then when he was freed, in 1945, he sat down for just nine days and wrote *Man's Search for Meaning*. If you have not read it, put my book down now and get a copy. At one level, it is a heart-stopping description of life in the concentration camps. At another, it is a luminous account of Frankl's unique thinking on the human condition.

According to Frankl, we do not want anything as simple as happiness. What we want is a sense of meaning – of what our lives are about. He suggested three sources for meaning. One source has to do with the people around us – what we mean to them, what they mean to us, what we can do for them, what perhaps we can do against them. The second source has to do with challenge and achievement – the sense that we have done what was needed and done it well. Frankl's third source was suffering, and it is easy to imagine that a man who had just emerged from Auschwitz would understand suffering in a way that most of us never will. Still, I think we can get a sense of what he meant. When we have lived through suffering, there is a sense that we have come face to face with reality, with both its good and its bad – that we have seen something important about both the world and ourselves.

Frankl was no ethologist but, in my opinion, his idea of meaning fits well with Lorenz and the IRM. The sense of meaning, I think, comes with discharge of the IRMs built into Pan and, just as Frankl thought, they do not work in terms of abstract 'happiness'. They are the concrete IRMs of Lorenz – we derive meaning from a moment of sexual joy, or from the smile of a child, or from the appreciative nod of a colleague, or from the ferocity of a last-minute goal that defeats a rival football team. In line with Frankl's thinking, IRMs are deeply concerned with other people; with effort, ambition and achievement (as I shall discuss in Chapter 11); and, quite possibly, with meeting and learning from all manner of challenges and suffering. They give us our sense of what our lives are for and why we are here. Spock can tell us how to do things, but only Pan can give us meaning.

CHAPTER 3

The idea machine

A vast store of knowledge

Of course, animals cannot be all innate releasing mechanisms and instinct. It is feasible for fixed patterns of behaviour concerning fixed aspects of the animal's world to be built into the genetic code, but where the world is variable, this cannot work. Instead, we need learning, often combined with instinct, so that the fixed and the variable can come together to determine the animal's behaviour.

Learning produces an internal model of the world, which the animal can then use in behavioural choices. The classic example comes from the psychologist Edward Tolman, studying how rats create internal models of a maze.[1] In the first stage of the experiment, the animal learns to run through a maze to reach some food. The route is not very direct but, for now, this is the only route available. Now he is put back at the start, he knows the route he has learned, but suddenly this route is blocked, and instead several more have been opened. We know what we would do, and very often the rat does just the same: among his new options, he chooses the one that leads most directly to the food. He has not just learned the one route he has experienced – he has learned the broader layout of the world and can navigate a new path from start to goal.

Slightly more complex, a cat sitting in the kitchen, seeing a rival cat walk past the house, runs straight round to the window of the living room to check on his rival's progress to the front. A wasp leaving its nest circles the opening, imprinting landmarks in its nervous system, and returning later it knows where to enter.[2]

Animals vary widely in the complexity of the models they can use, and psychologists differ strongly on exactly how this works, but throughout the animal kingdom, wherever the world is too variable to allow fixed action patterns, instinct is supplemented by learning.

In human beings, evidently, this act of building world models creates enormous banks of knowledge. We do not just know where to enter the nest; we have models of the atom, the big bang, what our children like for dinner and how to cook it, geometry and algebra, how to ask for a coffee, and on and on and on. We are like the rat, but raised to the nth degree.

I once attended a talk by an old friend, Tom Landauer, who tried to estimate the number of bits of knowledge that a human being acquires in their lifetime. He used several methods, such as estimating the rate of learning per minute and multiplying by the number of minutes in a lifetime, and after his talk another, rather cynical friend, Frank McKenna, described these methods as 'think of a small number, think of a very big number, then multiply them together'. However we calculate this number, though, it is going to be staggeringly large. We drink in information at a spectacular rate, and keep on doing it for millions of minutes.

Our knowledge comes in two broad forms. First, we have what psychologists, following the memory researcher Endel Tulving, call 'episodic memory'.[3] This is our memory for the events of our lives – the time we turned left at a roundabout and got lost, the time we cooked lasagne for a friend and it turned out she had just given up meat, our breakfast this morning and what our English teacher said to us when we were 12. Though each memory is a snapshot (or, probably better, a film), describing just one event from the past, it can be useful again in the future. Next time we come to that roundabout, we remember what went wrong and turn right instead; next time we think of saying 'due to', we remember our English teacher and, if his rule applies, substitute 'owing to' instead.

The second form of memory, however, is even more generalisable and useful. Tulving called it 'semantic memory' – our vast storehouse of facts and rules telling us in general how the world works. Semantic memory tells us that a tiger has stripes, that two and two make four, that lasagne can also be made just with vegetables. It is striking, though not central to this book, that different kinds of brain damage are needed to destroy episodic and semantic memory; one person can no longer remember that they spoke to you a few minutes before, while another can no longer tell you whether a tiger or a lion has stripes.

Now comes the question. We have filled the person's head with a staggering volume of knowledge, but how can that knowledge be used? Now, the person wants to do something – perhaps travel to a conference in Japan. The knowledge of how to do this is in there, but how does the person get it out?

Using the knowledge: Divide and conquer

Perhaps most obviously, our method for solving complex problems is to divide and conquer. We cannot say that we want to travel to Japan, and immediately ask how to move our body to make it happen. On its own, the high-level goal is not enough to decide whether, for example, we should move our left or right arm. The complex, whole problem has to be broken down into simpler, more specific parts.

Often, this is seen as a process of creating a structure of sub-goals beneath the main goal at the top. For example, we know that we want to travel to Japan and, consulting our enormous body of world knowledge, we set a sub-goal of taking a flight. Once we are dealing with the sub-goal, we can forget many other aspects of the whole problem – whether to rent a car on arrival, how to read road signs and so on. Instead, we focus just on flying, and now this brings up a new sub-goal – buying a ticket. For now, we can forget about what to pack and how to get to the airport – we are just focused on buying a ticket, and now we set the sub-goal of logging on to the internet. So far the relevant information has all been brought up from memory, but now, information from our senses comes into play. We are sitting at our desk – the computer is in front of us – we see the mouse. Now, at last we do know what to do with our body – we reach out our right hand, grab the mouse and slide the cursor across the screen.

This divide and conquer strategy is very powerful. Once we have seen it in action for flying to Japan, we can see that we do it all day every day. The earliest AI programs did the same thing to solve problems in formal logic – they had a large body of knowledge concerning the rules of logic, they had a statement of the problem and a goal they were to achieve, and by breaking the problem into parts, they moved from start to goal.[4] Their protocols as they did this looked remarkably like the protocols of university students solving the same problems.

An analysis of standard 'intelligence tests' points up this same ability to divide a complex whole into useful parts. For approaching a century, it has been known that simple puzzles, like those you might find in a book of brain teasers, have a quite remarkable property. People who do well with these puzzles also tend to do well in many aspects of life – in academic achievement, success at work, salary, even life expectancy.[5] The puzzles themselves are trivial but, obviously, the ability to solve them shows something broad and important about a person's mind.

An example is shown in Figure 1. There are four boxes arranged in a 2 x 2 matrix, three filled and the fourth empty. Based on the three that are filled, and the relationships you can see in rows and columns, you are supposed to work out what the missing shape in the fourth box should look like, and draw it in the separate response box under the matrix.

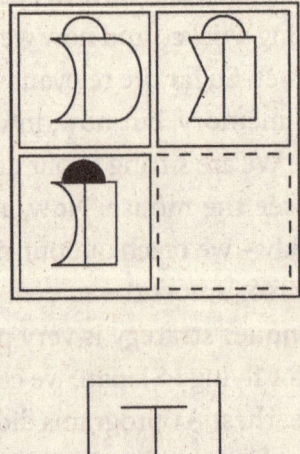

Figure 1. The task is to work out what shape should go into the dotted empty box, and draw it on to the framework given in the response box at the bottom.

To succeed, you will likely proceed by divide and conquer. You may focus on the right side of each shape – now looking at the pattern in the whole matrix, you conclude that, to get the best overall balance, the right side of the missing shape should be curved rather than straight. This way each row has one curved and one straight, and the same for the columns. You draw the curve into the response box. You focus on the top, conclude that the answer needs a little black semicircular hat, and draw that in. Finally, you focus on the left side, conclude that the left side of the missing shape should be angled, and the problem is solved. You can draw in the last part of the answer and you are done.

Explained this way it seems easy, but for many ordinary people it is not. A few years ago, in an experiment with problems exactly like this, my colleagues and I found the solutions of many people to be filled with mistakes.[6] The same thing happens in any ordinary intelligence test – and the people who struggled with our problems were usually the same people who struggled with other tests.

Once you have focused on each part of the problem in Figure 1, the answer seems fairly obvious, and to prove this, we can separate the complex shapes in each box into their parts. Now look at the separated versions in Figure 2. The problem is exactly the same, but you no longer have to divide it into parts – this has been done for you. You focus on the left matrix, decide that the right part of the answer is curved, draw it in; focus on the middle, decide the top should be black, draw it in; focus on the right, decide the left side should be angled, draw it in. The sequence of moves is the same, but now you did not have to create the sequence yourself. Now, every step seems truly trivial – you would hardly call it problem-solving at all – and indeed, now everybody in our experiment did very well, even those people who had struggled with the version in Figure 1.

Simple puzzles like these measure something important, and this something seems to be the ability to divide and conquer. All day every day, we divide complex problems into simpler parts, focusing on one part after another until the whole problem is solved.

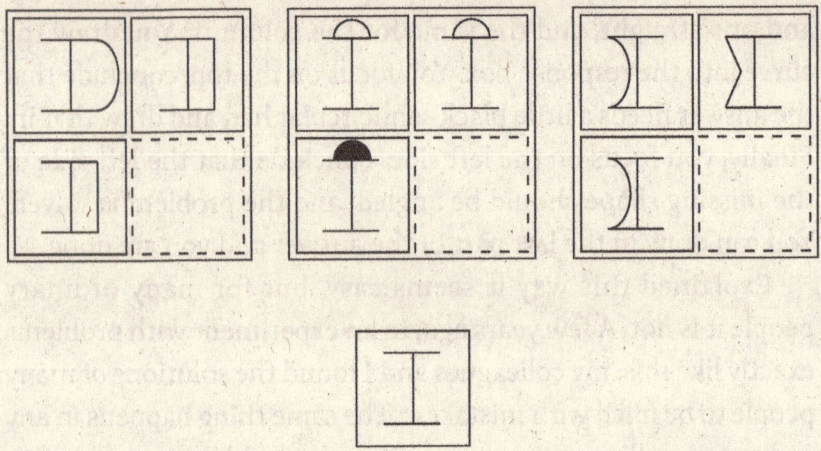

Figure 2. The problem is the same as Figure 1, but now divided into parts.

Creating simple ideas

What does it mean to focus on one part of the problem? It may seem simple, but with a little thought we see that it is not. Take just the single, easy step at the left of Figure 2. To solve even this simple problem, we need to appreciate the shapes in each box. Each shape needs to be linked to its correct position in the matrix. The pattern of three shapes needs to be connected to some conception of the 'rules' of the puzzle – that the whole thing when finished should look balanced and complete, with each changeable element appearing twice in the four boxes. Mild perhaps, but hidden in the background, there is also an emotional side – the 'aha!' sense when the answer seems correct and is ready to be written in. What we have done is build a little information structure in our mind – something that connects multiple parts with just the right roles and relationships, including, for example, 'the top right shape is straight', 'the missing shape should be curved', 'curved can be drawn into the answer box'.

Again, this ability to combine parts into new, structured wholes is everywhere in our mental lives. Most conspicuously, perhaps, it is the basis for the infinitely varied and productive world of language. If we have the concepts of John, Mary and hitting, and the words for these concepts, we can generate 'John hits Mary', or with the roles reversed, 'Mary hits John', or just 'John is hit', and so on and on. Our mental lives are filled with these small bits of computer code, each describing the roles and relationships of a cluster of parts.

And now we get to the really powerful bit. The parts, the roles and the relationships can be anything we can imagine. The machine for generating these bits of code can produce any idea, good or bad. It can be, 'If we build the arch to a point it will hold up the stones above' – and when we are constructing our church, this little information structure can be used whenever we come to a new window, with the rest of the project for now in abeyance. It can be, 'If we put the stone on logs, we can roll it along.' It can be an arbitrary rule, determined by nothing but social convention: 'The knight goes one step straight and one diagonally.' Of especial importance for this book, it can be an idea that riffs off the promptings of Pan. It can be, 'Our generals will wear epaulettes,' or more threateningly, 'The women will cover their faces.' The machine simply equips us with the ability to create ideas. The ideas themselves can be anything.

Whatever its content, a simple, focused thought is necessarily also an abstract thought. 'Abstract thought' is often taken to be the crowning achievement of the human mind. We admire the ability to see the geometrical principles of Euclid, or to conceive of justice and truth. Indeed, abstraction is the essence of Spock and the idea machine, but there is abstraction in everything he does, not just in the higher reaches of mathematics or law.

The entire point of divide and conquer is that a part of the problem can be solved while everything else is left to one side. As we work on buying an air ticket, it does not matter that the entire plan will need many other parts – the luggage, the travel to the airport, the arrangements made for feeding the cat. We are working on one property of the whole solution, while leaving everything else unspecified. Stating one property, leaving everything else unspecified, is exactly what we mean by 'abstraction'. Something 'fair' could be an agreement in the playground, an allocation of money or the resolution of an international dispute. The same property of 'fairness' is seen in many contexts, just as the same property of buying an air ticket can be built into many different plans.

Of course, some ideas seem more abstract than others. 'Fairness' leaves very many details of a particular social exchange unspecified, while 'redness' may be less widespread in its application. (Though having said this, it is hard to know exactly how any comparison like this can be made.) We may feel that Euclid is abstract while buying an air ticket is not, but fundamentally, I think, they are the same. In both cases, the idea machine generates something focused, capturing one aspect of a complex whole and able to work just with that one aspect, leaving everything else in the background.

Good and bad ideas

In *Star Trek*, Spock is the essence of reason. His understanding is all embracing, and his deductions based on this understanding are faultless. This indeed is one way that we can use our idea machine, and when we use it like this, we have the human 'reason' so valued through the ages.

The essence of valid reasoning is that, as we move from one idea to the next, we do so in a way that exactly reflects our knowledge of the world. We know that the first idea does indeed imply the second, and therefore that, if we know the first, we can infer the second. We know that unsupported weights always fall, and that if we pour boiling oil from the turret, the soldiers at the castle gate will be burned. We know that oil can be heated on the fire, and if we want to burn the soldiers, we can put the oil in the cauldron. We know that a fire can be made from wood, and if we need to heat the oil, we can carry over the faggots.

These chains of reasoning fill our lives, from defending the castle to moving through the steps of a geometrical proof. Indeed, I would argue, much of the power of mathematics comes from the perfection of its implication rules. When we see 'side angle side', we do not just know that the triangles are probably congruent, we know with 100 per cent certainty that they are. Because certainties are 100 per cent, we can string together mathematical arguments into indefinitely long chains, and providing we use the rules correctly, we still know that the conclusions at the end will be certainly correct . . . even when they have taken us into the impossible-seeming worlds of contemporary physics and astronomy, with properties we can barely comprehend.

The idea machine can be used for the ideal reasoning of Spock or mathematics but, as we all know to our cost, very often it is not. New ideas arise not because they have been carefully constructed from what we know to be true, but by much looser means. Very often we believe something just because it sounds roughly right, or just because we heard it from somebody that we trust. Nothing in our lives could work if we always demanded perfect deduction, and in the complexity of our world, many of our most precious ideas are demonstrably wrong.

One of my favourite examples is religious belief, provably

wrong in billions of individuals by the following simple argument. We have Muslims, Christians, Hindus, Buddhists and innumerable others. As all these groups believe different things, often to the level of quite specific details, at most one group could actually be right. (Of course, the thought leads to by far the most likely conclusion that all groups are wrong.) Later, we shall be looking at how big religions combine the promptings of Pan with an elaborate complex of ideas and rules added on by Spock. All these detailed ideas have been picked up from each person's surrounding culture, and for many reasons it is essential that we do this . . . but it is certainly no royal road to the truth.

In each thought, we use only a tiny amount of what we know

Certainly the abstract idea machine is one quintessential part of humanity, and when it is used well, it takes us to places no other animal could go. Its powerful properties, however, can also take us to quite the wrong places. In particular, the very strength of Spock – the ability to focus and abstract – is also his core weakness. Like drink, focus and abstraction are excellent servants but dangerous masters. If we stop to think, we realise that our mental lives are filled with ideas that, looked at a different way, could not possibly be right. We see just part of the problem, and many other sides – perhaps very well known to us – are for the moment left out of mind.

Let me begin with a simple example, introduced into experimental psychology by the German psychologist Karl Duncker.[7] As I mentioned in the Introduction, Duncker was interested in the mental processes of solving novel problems, such as how to use X-rays to kill a tumour without damaging the surrounding tissue.

In the course of his work, he encountered a striking phenomenon that he called 'functional fixedness'. The essence of functional fixedness is that we focus on the familiar use of an object, without seeing obvious possibilities for using it in a different way. In a sense, we of course 'know' the information we need, but as we consider the problem, this information does not come to mind.

Duncker's famous example of functional fixedness is the candle problem. A person is given a candle, matches and a box of drawing pins. He is asked to fix the candle to the wall and light it so that the wax doesn't drip. A common strategy is to try melting some wax and sticking the candle to the wall. Another is to try pinning the candle in place. Neither of these works, but the correct solution is often hard to see – take the drawing pins out of the box, pin the box to the wall, and stand the candle in it.

Another striking example comes when we try to follow a set of instructions. The instructions go in, and if the person is asked what he should do, he answers correctly. Then he tries the task itself . . . and the information promptly vanishes.

As first shown by a great Russian psychologist, Aleksandr Luria, this can be especially striking in people with major brain damage, especially damage to the frontal lobes.[8] In this case, the knowledge and the behaviour can be incongruent even in the same moment. The person is asked to lift up their hand when a torch is switched on. The examining psychologist switches on the torch, the patient says, 'I should raise my hand' . . . and nothing happens. This is not owing to any difficulty with the movement itself (though that can also happen after damage in other regions of the brain). If the psychologist insists, now the hand goes up. It is more like chronic absentmindedness – the patient somehow fails to notice that he has not done what he should.

I have worked on this phenomenon for many years, sometimes in patients with brain damage, but more commonly in perfectly healthy people.[9] Now, it is hard to make the words and the actions clash in a single moment, but it is easy to see it happen with just a few seconds' separation.

The person is given a few simple instructions, such as 'Watch a stream of letters and numbers coming up on the screen. Repeat letters as soon as they appear. Add pairs of numbers and say the result. When you see a plus sign, switch to a different stream on the opposite side of the screen.'

The person repeats all the rules, the stream starts, the letters are repeated, the numbers are added. Then the plus sign is shown, and the person just continues with the original stream, apparently undisturbed, as if nothing had happened. A couple of seconds later, the stream stops, and if the person is asked what they saw, they will often say 'a plus'. If they are asked what they should have done, they say 'switch to the other side'. They knew the rules, they saw the plus sign, but in the moment they just went on as if this part of the instructions had never been given.

Interestingly, the people who do this tend quite strongly to be the same people who struggle with matrix problems like the one in Figure 1. Just like Luria's patients, furthermore, these people are perfectly able to obey the instruction if they are given a few reminders. Again, it seems to be a problem of attending to the right thing at the right time – the knowledge is in there, but not brought out when it is needed.

Probably, under the right conditions, we have all experienced this. We are playing Uno, we know what to say when we are down to one card, we know we will be punished if we do not say it, we play our second to last card and with delight our adversaries shout 'Uno!' We ruefully take the two penalty cards . . . the rule was in there but, in the moment, our mind was elsewhere.

Attentional focus can produce all sorts of surprising phenomena. The sheer irrationality of much human thinking was revealed by a great pair of psychologists, Danny Kahneman and Amos Tversky. For his founding role in behavioural economics, overturning an economics based purely on rational calculation, Kahneman won the Nobel Prize in 2002, which doubtless Tversky would have shared but for his early death in 1996.

Here is one of Tversky's demonstrations.[10] Participants in an experiment are told to choose between two holiday destinations. The price is good on both, but the 'travel brochure' (this was 1993) gives only limited information. For one destination, the information is rather neutral and uninformative – 'average beaches', 'medium-temperature water' and so on. For the other, there is much more information, but it points in both directions – 'gorgeous beaches and coral reefs', 'very cold water' and more. In a triumph for optimism, 67 per cent of people choose the second option, apparently thinking that the great points will make up for the bad ones.

Now, a second group of people are asked about the exact same options, but with a negative slant. This time the participant has reservations on both options, and has to decide which one to cancel. Now only 52 per cent of people hold on to the gorgeous beaches and very cold water. Thinking of cancellation rather than purchase, negative features come more strongly to mind than positive, and the threat of cold water begins to dominate thinking.

The work of Kahneman and Tversky is filled with demonstrations that choices between the exact same alternatives are altered by the way that the problem is framed, quite contrary to ideal reasoning. In developments of this work, psychologists turned to people's ideas about themselves, their happiness and their life satisfaction, and in addition to the usual effects of attentional focusing, the results show that, often, we really do not understand ourselves.

Here is a simple and rather transparent example.[11] College students are asked, 'How happy are you with your life in general?' Having answered, they are asked, 'How many dates did you have last month?' The answers to the two questions are entirely uncorrelated – how many dates the person went on seems to have had no effect whatsoever on happiness.

Now, the questions are simply reversed – the person first answers about dates, then about happiness. Now, there is a huge effect – the judgement of happiness is very closely linked to the number of dates. What has happened? The explanation, surely, is that asking about dates brought dates to mind, and now, when the person turns to happiness, the dates are there in the foreground, dominating the judgement. Kahneman called this a 'focusing illusion'.[12] David Schkade and Kahneman summed this up with: 'Nothing in life is quite as important as you think it is while you are thinking about it.'[13]

This is already striking, but further work led Kahneman to a rather astonishing insight into our own failure to know ourselves. In these experiments, happiness was assessed in two quite different ways. In the first way, perhaps the most natural, people simply make an abstract judgement of how satisfied they are with their lives. With this method, the rating of happiness is predicted by some of the things you might think are most obvious – income, employment and (negatively) body mass index.

Now comes the second way – several times each day, the person rates how happy they are right now, and an overall measure is obtained by combining all the ratings. This method, ostensibly closer to the way people actually feel as they live their lives, gives quite different results. Now income, employment and body mass index have nothing to do with it. Instead, it becomes very important what you are doing right now. People hate commuting. They love being with their family.

Kahneman argues that, when people are asked to make an abstract judgement about life satisfaction, attention turns to the 'conventional' ideas of wealth and success.[14] These are not unimportant, but they are not what most of the moments in your day are about. And when we turn to major life decisions, it is the conventional ideas that tend to dominate. We think the new job will mean promotion, success and money but, arguably, if we want to be satisfied with our lives, it is the commute and the time away from the children that should really dictate our choice.

Spock's urge for consistency

In Chapter 1 we saw how Pan really does not care about consistency. Signals from the same conspecific can release aggression at one moment, courtship at a second moment and shared brood care at a third. Consistency is an idea of Spock's. With all his focus on one thing at a time, he likes to see that the ideas do not contradict one another. As new ideas follow one another, they are carefully selected to keep inconsistency to a minimum.

This idea was formalised in the 1950s by the psychologist Leon Festinger in his concept of cognitive dissonance. Festinger's experiments are filled with people who see only the arguments that support their own beliefs, or who think a product is more attractive once they have paid money for it, or that joining a society was worthwhile when it was hard to do. To listen to an argument against one's cherished belief, or to think that one paid good money for a poor product, brings a sense of dissonance, and Festinger proposed that people constantly adjust their thinking to keep dissonance to a minimum. As he famously put it, and as we all know full well: 'A man with a conviction is a hard man to change. Tell him you disagree and he turns away. Show him facts or figures and he questions your sources. Appeal to logic and he fails to see your point.'[15]

A similar urge for consistency appears in the moral reasoning studies of Jonathan Haidt.[16] In these studies, people were shown short vignettes carefully written to violate a common moral sense but to be demonstrably harmless. In one, for example, a woman cleans out her closet and finds an old American flag. Not wanting it any more, she cuts it up and uses the rags to clean her bathroom. Nobody sees her, it is just cloth, but the action violates a common sense of respect for the flag, and many people immediately judge it was wrong. Now, they are asked why and, commonly, they begin to invent fake victims, sometimes at odds with what they have explicitly been told. One person says a neighbour might see it and be offended – but the story states that nobody else knew. Rather winningly, Haidt recounts the explanation of a child – maybe the rags would clog up the toilet.

In Haidt's view, it is Pan who really makes the judgement, promoted by an instinct of loyalty to the tribe. Once the judgement is made, Spock tries to catch up by bringing consistent ideas to mind. In this context, he is a powerful rationaliser, thinking only what he wants to think.

We know this all too well, though, often, we see it clearly in others and miss it in ourselves. In March 2022, the Russian army had invaded Ukraine. The BBC ran the story of Oleksandra, hiding in her flat as Kharkiv was shelled, speaking regularly to her mother in Moscow.[17] 'I didn't want to scare my parents, but I started telling them directly that civilians and children are dying,' she says. 'But even though they worry about me, they still say it probably happens only by accident, that the Russian army would never target civilians. That it's Ukrainians who're killing their own people. It really scared me when my mum exactly quoted

Russian TV. They are just brainwashing people. And people trust them.'

Born in the 1950s, with the Second World War still fresh in the minds of my parents and teachers, I was brought up to mock the transparent propaganda that speaks of brave invading liberators freeing the people from their oppressors, but this propaganda exists because it works. It is painful to think that one's leaders and one's country are bitterly at fault, and, for all of us, more than easy to think that they are not.

Simple ideas in a complex world

The core difficulty for Spock is that reality is complicated. Life choices, political alternatives and moral issues have many sides to them, with no single truth to be found. This is a difficult matter for a mind that constructs simple ideas. Many of the ideas are in conflict, and our cherished beliefs, political slogans and moral precepts can never capture more than one side of complex reality.

A religious commitment tells us to love our neighbour. This is all very well, but when our neighbour keeps stealing our chickens, we are likely to conclude that love is not always all you need. When we debate politics we may feel that the will of the people should be sovereign, but if the will of the people challenges something we otherwise hold dear, perhaps sexual or religious freedom, the principle has to go.

Our days are filled with simple social rules and precepts, sometimes quite arbitrary. Children should be seen and not heard. Only the right hand must be used for eating. Milk and meat are to be cooked in separate utensils. Faced with the confusing challenges of noisy children or food hygiene, Spock invents simple rules, and if the child now speaks up or the cook mixes the pots, the consequences can be severe.

The complexity of reality means that, often, what we say or believe at one moment simply does not fit what we say or believe at the next. Americans commonly argue that their country is the land of opportunity, with a chance for anybody to become president. Yet if they have read *The Jungle*, Upton Sinclair's 1904 novel of new immigrants struggling and exploited in the Chicago meat-packing industry, or seen the heart breaking classrooms of the fourth season of *The Wire* (2006), they really know perfectly well that for many Americans opportunity is little to none; and for their own children, they move heaven and earth to assure all the educational benefits that wealthy circumstances can afford.

The Declaration of Independence opens,

> We hold these truths to be self-evident, that all men are created equal, that they are endowed by their Creator with certain unalienable Rights, that among these are Life, Liberty and the pursuit of Happiness . . . when a long train of abuses and usurpations, pursuing invariably the same Object evinces a design to reduce them under absolute Despotism, it is their right, it is their duty, to throw off such Government, and to provide new Guards for their future security.

It seems fair to assume that the writers knew full well that their slaves were men – but different issues surface when a person is writing a bold set of abstract rules, and then later managing affairs on his estate. At a dinner party, it would bring no surprise to hear the same friend state that stereotypes should be avoided, then later that Russians are callous or the British are dull. Each statement can easily be justified by considering just a part of complex reality, and often we move from one to the other, failing to notice that something must be wrong. At each moment, we bring to mind just a small, preferably consistent part of the truth.

This is perhaps the greatest limitation of Spock. With each abstract idea, he is trying to impose order on reality, but reality often has no simple order. Instead, it is a chaos of conflicting considerations, all competing to be recognised, often entailing opposite conclusions about what to believe or how to behave. Corresponding to the many sides of a question, Spock has constellations of many ideas, some that are compatible with one another, many that are not. When he is asked what to do, he makes a careful selection, turning the complexity of truth into a simplified version. For now, this simplified version is all he thinks of and, just for now, he is quite sure he is right.

The dialogue

Now we have the two players clearly on the stage – Pan, inflexible, fragmentary, inconsistent, but alert to the many subtleties and complexities of human interaction; Spock, abstract, powerful, focused, sometimes brilliant, often blinkered. Pan can think only of what he knows – how to play with a child, how to greet a friend, how to flirt or show respect to his mother. Spock can think of anything, but just one idea at a time, and it is hard for him to see that his clean, abstract ideas are always just one part of the complex truth.

Spock and Pan are in constant dialogue. In Chapter 2, we saw how, across cultures, common themes from Pan are filled in with many different details. In one culture, the shoulders are emphasised with an elaborate display of feathers; in another, with brocaded epaulettes; Pan provides the broad structure, but the ideas of Spock, sewn together from the opportunities of a particular time and place, fill in the details.

In Chapter 1, we saw how rival IRMs struggle to control an animal's behaviour, with the weights of each competitor varying over time and context. The ideas of Spock also have weights, pushing thoughts and choices in different directions. Sometimes, Pan's choice is supported by Spock's abstractions, but sometimes the two voices are at war. Both Pan and Spock offer their fragments of mental computer code, competing to control what we will do and how we will think. Choosing what to do can be hard for a courting beetle or a battling stag. With the competing voices of Pan and Spock, no wonder it is so hard for us.

CHAPTER 4
Two heads

Two kinds of brain functions

Pan and Spock have very different requirements. Pan needs hardware to run stable programs. Though some are doubtless unique to *Homo sapiens*, there is also much in common with the programs of vertebrate ancestors that have existed for hundreds of millions of years. Spock in contrast needs the flexibility to write new programs with any material available to human thought. Spock's ideas can be anything – a cultural rule, a nursery rhyme for a child or a theory of quantum mechanics.

Lorenz had detailed ideas of how instinct would work in the brain. It needed a specific neural structure that could execute the behaviour of the innate releasing mechanism, driven by the appropriate releaser, with a need to execute that fluctuated with time. When Lorenz was working, these physiological conclusions were far from anything that could actually be measured, but with modern methods for investigating the brain, they have very much become reality. Specific structures in an animal's brain run the bits of code that control the IRM, and at the heart of this are ancient regions, arising far back in evolutionary time.

We know most about the neurophysiology of the IRM from other animals since, in the human brain, it is usually not possible to make the same detailed measurements we can take from flies or mice. For the idea machine, it is different, since no other animal has an idea machine remotely as powerful as ours. Again, however, recent decades have seen major developments, largely through new methods for imaging human brain activity. Here we are working at a scale much coarser than the scale of the fly brain, and, undoubtedly, we know much less. Already, however, we know that the neural hardware is something very different from the hardware of evolutionarily ancient behaviour and the IRM.

Inevitably, this chapter is rather unlike the rest of the book, with a whistle-stop introduction to neurobiological concepts and methods, and especially, the complex terminology needed to refer to different structures in the brain. With all this detail, I will try as far as I can to get across the general principles – the principles of how Pan and Spock are implemented at the level of brain operations.

Neurons

To understand modern work on the IRM, we need to begin with the nerve cell or neuron. Each neuron is a tiny computational unit, taking in information from many inputs and combining this information into some useful output. To show how tiny, consider that the human brain, fitting comfortably into the head, contains around 100 billion neurons. To understand the brain, we need to know how these enormous networks of neurons combine to turn information from the world into useful behaviour.

Inputs arrive at a neuron on a treelike structure of tiny threads known as dendrites. These threads converge on the main cell body, where inputs are combined and a result is calculated. Emerging from the cell body is a fibre, the axon, used to transmit the output of this neuron in a series of brief electrical impulses. These impulses in turn are received by the dendrites of the next neurons downstream, where again they can be combined with others. A classical example is the early processing of visual information in the nervous system of a cat.[1] Processing begins at the back of the eye, in the retina. Each retinal neuron receives inputs from the light receptors of the eye, constructing the message that a patch of light in a certain colour has been seen at a certain place in the visual image. When this patch is present in the scene, the neuron fires off a series of impulses, sending its message into the brain. When this message arrives, the next step is to combine inputs from multiple neurons; perhaps individually these inputs indicate a connected row of patches, and now they are combined by an integrating neuron to signal a line in a certain place, oriented at 60 degrees to the horizontal. This message is sent on again, now in the electrical impulses of this 'oriented line' neuron, and combined with the messages of many others, it might be assembled to identify the presence of a red triangle.

This is a very over-simplified picture. In fact, in each brain area there are neurons of many different types, each making its own contribution to the way this area works. There is not only excitation by inputs from upstream neurons, but inhibition between one neuron and another. For example, a neuron signalling a red line may inhibit another signalling a blue line, ensuring that the system's output tells a consistent story, and at least as important, preventing the whole system from running out of control with a barrage of activity everywhere. Other neurons bring in signals of sleep or wakefulness, or signals promoting selective attention or learning.

Neurons vary widely in the shape of their cell bodies, the pattern of their connections, the length of their axon and thus how far their message travels through the brain. At the end of the axon, each neuron transmits its message to the next by releasing a chemical signal or neurotransmitter, and many different types of neurons release many different neurotransmitters, each with its own effects.

Unravelling this complexity is a staggering challenge, whether it is the fruit fly *Drosophila* with something above a hundred thousand neurons stuffed into a brain the size of a poppy seed, or ourselves with 100 billion. Even for the simplest nervous systems, there is much that is still unknown. At the same time, recent years have brought spectacular new opportunities for measuring, changing and understanding the activity of large populations of neurons.

Much of this progress has been brought about by genetic engineering. To measure the activity of individual neurons, they can be genetically altered so that, when they fire an impulse, they emit a tiny pulse of light that can be picked up under a microscope. Now, with the microscope positioned over a structure in the brain of a fly or mouse, we can simultaneously record the activity of hundreds or thousands of neurons as the animal courts a mate or protects its young. Alternatively, the neuron can be made sensitive to light; when a laser is turned on, the neurons that have been tagged can all be switched on or off, showing how their outputs influence other neurons in the circuit, or perhaps the behaviour of the whole animal. We can genetically engineer different classes of neurons in a brain region – for example, just the inhibitory neurons, or just the neurons whose axons depart for a particular target region, or just those that release a particular neurotransmitter. It may be a laser that we use to change a neuron's function, but other methods use a change in temperature or the targeted response to a drug. Limited though our understanding certainly is, with these methods we begin to see where and how the brain runs its IRMs.

Innate releasing mechanisms in *Drosophila*

The fruit fly *Drosophila* is a good starting point, simply because we know so much. The insect brain is very different from the brain of a vertebrate. In each segment of an insect's body, ganglia (or collections of nerve cells) give a degree of control over the actions of that segment. At the head, several segments of the insect's body are fused together, and the fused ganglia form what 'brain' the animal has. A large research enterprise has now identified every cell and every connection,[2] and staggering though the complexity already is, this is something that could not be imagined for our brain.

Drosophila, of course, has IRMs for many aspects of its behaviour, and to illustrate, we can take courting and attacking.[3] If a male fly is approached by a female, it picks up the female pheromone and begins to sing. If it is approached by a male, it picks up the male pheromone and begins to attack. There is much in the fly's brain that is needed to pick up and identify the pheromones, and to coordinate all the movements of singing to a female or lunging at a male, but at the heart of this IRM is a remarkably simple system.

In a set of neurons called the P1 cluster, with perhaps 20 neurons in each half of the fly's brain, we see a master controller for both courting and attacking. P1 neurons begin to fire when another fly is encountered – for example, when the male touches the body of a virgin female.[4] If these neurons are artificially activated, the fly begins either to sing or to lunge, with the choice determined by the exact details of the stimulation, or by the presence of a critical pheromone. If neurons are activated when the fly is alone, nothing happens, but a few minutes later, when the fly is given access to another male, it immediately starts to attack. The artificial stimulation left him ready to pick a fight as soon as the opportunity arose. Activity in P1 neurons shows just the fluctuations over time that Lorenz anticipated. For example, excitability is high if the animal has not recently mated, then diminishes with sexual satiety.

Further work asks how pheromone information reaches the P1 cluster from the fly's chemical sensors,[5] and how commands are sent to neurons in lower segments of the animal's body, directly controlling its movements of courtship or attack. With many details still to be filled in, we have the outline of an entire IRM, complete with releaser, fixed action pattern and, when conditions are right, its own spontaneous need to discharge.

Mammalian Pan

The brain of a mammal is very different from the brain of a fruit fly. Mammals have an enormous cerebral cortex, evolved from a structure called the pallium in earlier vertebrates but now massively expanded, especially in humans. Lying beneath this cerebral cortex, however, are many other structures or 'nuclei', preserved relatively unchanged from fish and amphibians to ourselves. In these 'subcortical' nuclei, we find much of the hardware for the IRMs of courtship, aggression, parenting, defence – with variations from one animal to another, but with strong common themes.

Among mammals, mice are the equivalent of *Drosophila*, with many of the same opportunities for genetic engineering and resulting powerful experiments on neural populations. We can take aggression as an example. In the mouse, a group of subcortical nuclei with exotic names form a 'core aggression circuit'.[6] Salient members of the circuit are the ventromedial hypothalamus, the amygdala, the ventral premammillary nucleus – all evolutionarily ancient, and all with properties resembling what we see in the attack functions of the *Drosophila* P1 cluster.

In the ventromedial hypothalamus, the critical cluster now consists of perhaps 2,000 neurons – still a tiny fraction of the mouse brain. These neurons become active when the mouse decides to attack, and driving them artificially has little effect in an isolated animal, but can start a fight if two males are together. As one male sniffs another, the rate of firing in these neurons predicts whether or not the sniffing will move on to attack. Just as we saw in *Drosophila*, stimulating these neurons artificially puts the animal in the mood for a fight. If it knows that pressing a lever will give it access to another male, when the critical neurons in its hypothalamus are stimulated, it is quicker to press the lever and reach this potential victim.

Structures of this core aggression circuit are common to many vertebrates but, of course, the releasers that reach this circuit to trigger aggression vary widely from one animal to another. In mice, there is a strong dominance of odours from a rival male. In songbirds, instead, it is the rival's sound that triggers aggression, and now there are strong inputs to the ventromedial hypothalamus and amygdala from brain regions that process sounds.

Similarly, the outputs from this aggression circuit must control the detailed, stereotyped movements of many different aggressive behaviours. In the mouse, prominent outputs go through a region called the periaqueductal grey, whose neurons project into the spinal cord to drive the coordinated muscle contractions that make up a pattern of movements.

The periaqueductal grey is involved in the movements of multiple IRMs – the biting and vocalisations of aggression, but also the stereotyped grooming of young mouse pups.[7] In songbirds, an equivalent structure is active as a young bird learns to produce its species-typical song.[8]

The core aggression circuit is matched by similar circuits underlying different kinds of IRM. Mating and courtship also involve the ventromedial hypothalamus, but also another nearby nucleus of the hypothalamus, the medial preoptic nucleus.[9] This same medial preoptic nucleus is also centrally involved in the IRMs of parenting, including nest building, grooming of the pups and retrieving them if they are lost – with distinct groups of neurons involved in each behaviour. The anterior hypothalamus is important in defence.[10] All these behaviours are complex, with distinct neural populations for their different components – detecting the appropriate target individual, approaching and carrying out the final act.

Hormones

The hypothalamus and associated subcortical structures are not only ancient. They are also the site for hormonal and other influences from outside the nervous system, also shared by many vertebrates. Often, these influences play a key role as the strength of an IRM varies over time and circumstances.[11]

In a hungry animal, the empty stomach releases a peptide called ghrelin. Ghrelin circulates in the bloodstream and, in the ventromedial hypothalamus, binds to chemical receptors on aggression-promoting cells. The animal is ready for a fight – indeed, if ghrelin is infused directly into this region, the animal picks on another male and attacks. I do not know much about ghrelin in my own body, but I do know that when they were small, my children were always especially careful not to annoy me just before lunch.

Sex hormones modulate many aspects of behaviour, and receptors for detecting these hormones are richly distributed in the hypothalamus and other subcortical structures. In many animals, for example, aggression is related to the level of testosterone, matching sensitivity to testosterone in neurons of the medial hypothalamus. In another subcortical nucleus, the nucleus accumbens, the female sex hormone oestradiol changes neural excitability, and when oestradiol increases, the female is more ready to approach a male. In the medial preoptic nucleus of the hypothalamus, so critical to parenting, neurons are highly sensitive to hormonal changes over pregnancy. Furthermore, the route from brain to hormones is two-way. Through its close relations with the pituitary gland, the hypothalamus is involved with the synthesis of many hormones and their release into the bloodstream. Aggression, parenting or sex are not just a matter of the nervous system but are matched by corresponding changes in hormones with influences throughout the body.

There are also other slow changes in the strength of IRMs, again mediated by ancient subcortical structures. Rodents show highest aggression at the start of the night, and something similar may happen in Alzheimer's patients, who are most aggressive in early evening. These kinds of circadian rhythms are heavily controlled by another region of the hypothalamus, the suprachiasmatic nucleus, and, in mice, signals from this nucleus inhibit the core aggression circuit when the day begins, then release this inhibition with the onset of night.

Human innate releasing mechanisms: Adding in the cerebral cortex

Ancient structures like the hypothalamus play a central role in many forms of instinctive behaviour but, of course, they do not operate alone. From the sense organs that detect critical releasing stimuli to the organised motor systems that generate elaborate patterns of behaviour, IRMs depend on many parts of the nervous system. In particular, ancient brain structures interact with the more recently evolved cerebral cortex in mammals, or the equivalent pallium in birds. Several nuclei in the pallium, for example, allow the learning and production of birdsong, with its key role in competitive territory defence. In mammals, signals of all kinds reach subcortical nuclei in part after processing in the cerebral cortex, doubtless adding many levels of nuance and control to the IRM.

The human being has all the same subcortical nuclei as a mouse – the hypothalamus divided into its multiple nuclei, the amygdala, the periaqueductal grey and the others – and the same circulating hormones affecting neural function. We well know that, like mice, we too are up for a fight when we are hungry, or when we see and hear a jeering mass of rival supporters at a football match. In some cases, brain imaging studies confirm a role of subcortical structures in simple, IRM-like behaviour. One example is the role of the periaqueductal grey in predator avoidance. In a virtual pursuit game, the person controls a dot running through a maze, pursued by another shape representing a predator. If the person's dot is caught, there will be a painful electric shock, and when the predator suddenly looms close, there is a sense of dread and a burst of activity in the periaqueductal grey.[12]

Mostly, however, studies of the human brain focus on the cerebral cortex. This is largely a matter of technical limitations. Over the past few decades, the major method for localising activity in the human brain has been magnetic resonance imaging (MRI), and compared with what we can do in flies and mice, this method has very low resolution. In a typical 'functional MRI' study, the brain is divided into small cubes, about the size of a peppercorn, and activity is separately measured in every one of these cubes. Each cube, however, could contain more than a million neurons, each with its own firing pattern, and to limit things even further, MRI does not measure the electrical pulses of neurons themselves. Instead, it measures changes in the state of the blood when a region of the brain becomes active. The technique is immensely valuable, but is limited to what it can see and, in particular, it is more suited to measurements in the large cerebral cortex than in small subcortical regions such as the nuclei of the hypothalamus.

In the cerebral cortex, brain imaging studies of social functions show something striking. Social understanding is something very special, with its own dedicated brain regions that become active whenever we are thinking about the minds and actions of other people.[13] This social processing network can be activated in many different ways. You can play stories asking how people thought and what they intended to do, comparing them with control stories involving only inanimate objects. You can show moving shapes on a computer screen, comparing random movements with the strong perception of social interaction that arises, for example, if one shape appears to chase the other. You can ask which brain regions turn on together when people interact in a film. All these methods produce the same social processing network, with distinct nodes in different regions of the cerebral cortex, and close communication between them. Some of these nodes lie close to other cortical areas important in the sensory analysis of social stimuli, such as faces and bodies.

The functions of the different nodes in the social processing network are under intensive study and, at present, it is fair to say that little is really understood in any detail. What is clear, however, is that these regions specialise in social thinking, and as they analyse the thoughts, beliefs and actions of other people, surely they will add many layers of complexity to the subcortical management of social interactions.

Even for a fighting fish on the reef, there is much more than the bright colour of a rival that goes into the decision whether to attack or flee. In our case, a singing group of drunken young men spills out on to the street ahead of us. A stranger on a foreign street approaches and asks for change. Halfway down the ski slope, your young son races by yelling, 'Tally ho!' – but you are not quite sure whether he has seen the sudden drop before the trees below.

It is unlikely that, on its own, a subcortical nucleus can navigate the complexities of these social situations. A powerful cerebral cortex has evolved to interface with the brain circuits that came before it and, working together, they create our elaborate, sometimes crude, sometimes finely differentiated social lives.

A cortical circuit for Spock

Now, we can think about Spock, and his free generation of any idea. What brain system do we need for him?

Here, we cannot be thinking of small nuclei and counting cells in thousands. Any idea means integrating contents from across the brain, including our very large cerebral cortex. It has been known for more than a hundred years that different parts of the cerebral cortex are specialised for different things. Visual information first arrives at the back of the brain, in the occipital lobe, and is then processed through an elaborate series of 'visual areas', identifying movements, faces, objects, buildings, other people's bodies, where a cup lies with respect to our eyes and hands, and how we should move to look directly towards it and pick it up.

The same is true for other sensory systems, with auditory areas moving from an analysis of pitch and timing to recognition of speech, or areas that process inputs from the vestibular system and tell you how to stand up straight. Large sections of the left hemisphere are devoted to understanding and producing speech; similar regions, more salient in the right hemisphere, process social relations; a large region of the frontal lobe, on the medial surface where the two hemispheres lie up against one another, seems especially concerned with thoughts about ourselves; lower in the frontal lobes, on the surface lying above the eye sockets, there is regulation of mood and personality; large areas control movements, especially movements of the opposite side of the body, leading to the well-known hemiparalysis when they are damaged by a stroke.

If we scan a person watching a film, the entire cerebral cortex can be divided up into networks, some regions responding whenever there is rapid movement, some when we watch a person manipulating an object, some when events are puzzling or tense, some when people are talking.[14] This can be done to any depth – you can get 10 large networks, each consisting of regions whose responses are broadly similar, or divide them further into 20 smaller ones, and even with many more than this, the picture is still making sense, with consistent differences between the exact events that drive one network or a similar one close by.

When Spock creates ideas, he must be able to draw content from any of these networks, in any region of the cortex. Furthermore, he must be able to combine them in exactly the right way. He needs to know that he will pick up the hammer, select the correct nail and hold it to the wood, and hit the nail until it is in. It is no good picking up the nail with the wrong hand, or holding the hammer against the wood, or selecting a nail so big that this piece of wood will split. Then he needs to know that his workmate has stopped for coffee and can be asked to put more water in the kettle.

I have spent a good bit of my life working on brain mechanisms for this sort of process, and it would be an exaggeration to say that we understand in any great depth. Still, we are beginning to see at least an outline sketch.

A first clue comes from the 'intelligence tests' we considered in the last chapter. As we already argued, simple puzzles, such as completing the missing square in the matrix of Figure I, measure something important about our mental machinery. People who solve these puzzles easily tend to do well in many aspects of life. Now, we can use functional MRI to ask which regions of the brain house this machinery. For this experiment, we contrast problem-solving with a control task in which people look at similar figures and respond with similar movements (usually key-presses, which are convenient in the scanner) – but where the answer is obvious and there is no real problem to solve. What is the special extra ingredient added by the intelligence test?

The answer is intriguing.[15] On the one hand, several different brain regions are involved, broadly distributed across the major divisions of the cerebral cortex – the frontal lobe, the parietal lobe, the border of temporal and occipital lobes, including cortex on both outer and inner surfaces of the cerebral hemispheres. On the other hand, broadly distributed though they are, these regions are quite tightly defined. Right next to them are other cortical areas, with quite different properties such as response to language or importance in hand movements.

The second clue arrives at the same answer from a different direction. Problem-solving tests are important because they seem to concern cognitive mechanisms that contribute to much of what we do. So . . . are the critical brain regions also involved in many different tasks, whatever the content?

Over the past 30 years, brain imaging has been used to examine a huge variety of mental activities. There is problem-solving, but also keeping information in short-term memory, or trying to remember it for later, or deciding which of four lines is the longest, or trying to press keys in a random order, or moving the eyes away from instead of towards a flashing light, and on and on and on. When this technology was invented, the expectation was perhaps that each of these activities would have its own dedicated brain machinery, and as we have already seen, this is a major part of the truth. Soon, however, the results began to show something that had not been expected. As almost any task is made more complex or demanding, a part of the brain's response is activity in the same broad set of areas.[16] These regions, furthermore, overlap rather closely with the regions that are specifically active during problem-solving. An up-to-date version is shown in Figure 3. For obvious reasons, I call this the 'multiple-demand' or MD system.

With the powerful data of modern work, we see clearly that, in each of the two cerebral hemispheres, there are nine separate multiple-demand regions. They are broadly spread across both outer and inner surfaces of each hemisphere. We have already seen that the cerebral cortex interacts with more ancient, subcortical structures, and as the power of imaging data grows, we are also beginning to identify specific subcortical regions that behave in this same MD way. It is plausible to see this as a network that manages all kinds of cognitive activity, no matter the particular content. This is just what we need for Spock.

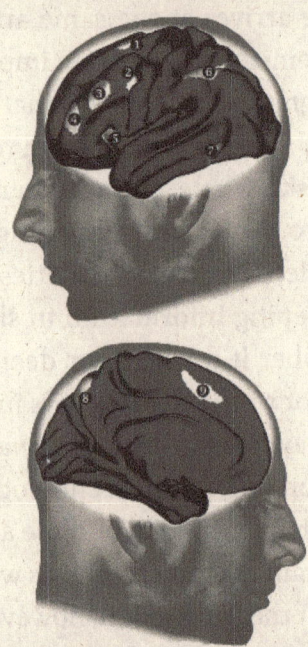

Figure 3. Pale regions show the nine regions of the multiple-demand system in the cerebral cortex. The network is illustrated on the left cerebral hemisphere, though similar activity also occurs on the right. The top figure shows the outer surface of the left hemisphere; the bottom shows the inner surface, which in the intact brain lies against the inner surface of the right hemisphere.

Why is it important that the parts of this MD network are so broadly spread across the cortex? In cerebral cortex, connections between one set of neurons and another are always strongest when they are short. The brain does not like to waste energy building long connections so, where possible, processing is organised so that regions needing to communicate are reasonably close together. Placing MD regions selectively across the cortex means that, one way or another, almost any cortical region, processing almost any content, has close access to its nearest MD entry point.

But whatever the content of an idea, its different parts

need to be put together into just the right combination. A second property of MD regions is perfect for this. Far apart though they are, different MD regions are strongly connected to one another.[17] We can see this connectivity by looking at their individual time-courses while a person watches a film or simply lies still in the scanner and rests. Whatever the person is doing, these widely separated MD regions increase and decrease their activity together, suggesting they are in constant communication. This strong connectivity means they should be able to put different bits of information together . . . just what Spock requires for his flexible generation of any sort of idea.

MD regions are especially concentrated in the frontal lobe of the brain. It is hard to move from the very precise regions defined by brain imaging to the consequences of damage from a disease such as a tumour or a stroke. Tumours and strokes are variable and messy, so it is not easy to identify exactly which regions of damage lead to particular behavioural difficulties. Broadly speaking, though, we have known since the nineteenth century that major damage to the frontal lobes leads to a striking pattern of cognitive disorganisation. The victim may be able to execute each isolated component of thought or behaviour, but somehow be unable to assemble these components into an organised, effective whole.

I have already mentioned the Russian psychologist Aleksandr Luria, whose work in the last century provided beautiful analyses of these difficulties.[18] He describes patients who want to poke the fire, but use a brush instead of the poker; who ring the bell for the doctor but, when he arrives, have no idea why they did it; who are asked to explain the meaning of a painting, and pick on just a few salient details to provide a confident but completely inaccurate analysis. Patients like these have often confused clinicians because they are perfectly able to walk, talk, see and remember, and indeed they are able to do many things in the carefully controlled setting of a psychological test. Then back in their daily lives, they are unable to cope, getting hopelessly lost as they struggle with multitasking at work or preparing all the dishes for the evening meal.

Of course, this is what we should expect if something important in Spock is broken. The simplest ideas may be fine, but as soon as there is a larger structure of thought to be created and used, this structure falls apart.

We might also expect that, if the MD system is important in a standard problem-solving test, then patients should struggle with these tests when the MD system is damaged. This also turns out to be true – whether it is in the frontal lobe or the parietal lobe, the total volume of damage to MD regions predicts how much a patient will have lost in a standard 'intelligence' test.[19]

Spock's neurons

Though no other animal generates ideas as we do, our MD system is likely to resemble similar systems in other animals, especially our closest relatives. In a monkey, we can record the activity of single neurons in the frontal lobe as the monkey carries out a task, usually something arbitrary that we have trained it to do with shapes on a computer screen and responses made with a lever, or simply by looking to one part or another of the display. Again, the way these neurons behave seems well matched to the requirements of a rather Spock-like machine.

In many regions of the brain, neurons fire in rather specific circumstances – as we saw before, perhaps when a certain shape appears on the screen, or when a rival male is about to be attacked, or when an animal makes a specific movement of his forelimb. Occasionally, single neurons can be recorded in the brains of human patients, on electrodes that have been put into the brain for some clinical purpose. Sometimes these recordings show astounding specificity – a neuron, for example, that responds just to the idea of Jennifer Aniston, whether it is her face, her name or even just somebody we have often seen with her in *Friends*.[20]

In the frontal lobe, however, the picture is often very different. Instead of responding to one fixed thing, neurons pick up *something that matters in this particular task*. If the monkey is making a decision between squares and triangles, a neuron may fire for squares and not for triangles; then in the next task, the decision is between a beetle and a goldfish, and now the neuron turns into a goldfish detector. What this looks like is a highly programmable system, coding the information that is needed to control whatever it is that the animal is doing.[21]

Often, furthermore, frontal neurons code not just one aspect of the task, but a combination of several things. Suppose that the monkey is remembering a series of three pictures, which it will need to choose in turn when it is tested a few seconds later. A neuron may respond to the picture of Jennifer Aniston – but only when she is the first picture in the series.[22] This sort of combination is critical for structured thought. Just as 'John hits Mary' is not the same as 'Mary hits John', the monkey's task cannot be solved by simply knowing that the series featured Jennifer Aniston, Lisa Kudrow and Courteney Cox. The system needs to know that 'Jennifer Aniston was first', 'Courteney Cox was second' and 'Lisa Kudrow was third', and to achieve this, it is very useful if single cells fire only for a particular combination of person and their place in the sequence.

Intriguingly, flexible neurons like these can also be found in the pallium of crows, often thought to be the most intelligent problem solvers among birds.[23] The idea machines of the monkey and the crow may be massively weaker than our own but, already, there is a system for novel, flexible management of arbitrary ideas, quite different from the fixed action patterns of the IRM.

Competing ideas

The MD system sews together the different parts of an idea, with their representation in different areas of the brain. But which idea? As we saw in Chapter 4, the workings of Spock are based on mental competition, with one idea allowed into the focus of attention and others for the moment suppressed.

Suppose you come down to the kitchen in the morning. If the first goal is coffee, you may find beans in the cupboard and take them to the grinder. The knowledge of what you need, the sight of the beans in the cupboard, the hand reaching out to pick up the bag, all these and other components of the process are coded across many separate brain areas. But suddenly the cat mews and you decide to feed her first. Now there is the sound of the mew, the sympathy for the cat's needs, the knowledge of where her bowl sits on the floor and the trip across the kitchen to reach it . . . everything concerning the coffee is suspended as the brain turns to all these components of the new choice.

It is now well known that, as attention shifts from coffee to cat, activity across much or all of the brain shifts accordingly.[24]

At the back of the cerebral cortex, where visual input arrives, neurons responding to the bag of coffee beans stop firing, and instead, new neurons indicate the tin of cat food sitting just next to the coffee beans on the shelf. The evidence comes from both single neurons in behaving animals, and from human brain imaging. It is not usually coffee and cat food, but the principle is clear. The animal moves his attention from a red bar to a green circle, and in the visual system, activity in 'red bar' neurons diminishes while activity in 'green circle' neurons appears.

Such findings suggest that different patterns of neural activity are in competition across many or most brain regions. If the visual system is responding to coffee beans, its response to the tin of cat food is suppressed; if attention switches to the cat food, this, in turn, suppresses the previous response to the coffee beans.

This happens everywhere in the brain, and very likely, it is coordinated by the MD system. All the parts of Spock's current idea are favoured; everything else is cut out. By this wholesale regulation of brain activity, Spock generates his focused ideas. Our experience of attentional focus is reflected in the attentional focus of the brain.

Spock's possible ideas compete; Pan's rival IRMs compete; and quite possibly, the two compete with one another. Suppose we use functional MRI to measure brain activity while a person lies in the scanner watching films. Over perhaps an hour, activity in each region of the brain fluctuates up and down as events in the story unfold. Very recently, in my research group, we have been comparing time-courses for amygdala and MD regions. As we saw earlier in the chapter, the amygdala is a key player in emotional responses and IRMs. The MD system is the key player in Spock's generation of novel ideas. And across the film, as one goes up, the other goes down, just as Pan and Spock compete for control of our minds.

Ancient and modern: Working together

I couldn't resist calling this chapter 'Two heads', since the relationship of Pan and Spock is well captured by the popular saying that two heads are better than one. Of course, we do not have two heads – the brain is an integrated whole, with ancient and modern structures collaborating to manage our thoughts and actions. Just the same, Pan does depend heavily on the ancient functions of hypothalamus, amygdala and more, with all their inconsistencies, temporal fluctuations and hormonal influences, and all their resemblance across animals from fish and frogs to birds and mammals. Spock, in contrast, needs machinery to construct arbitrary ideas from novel components, sewing together content from all over the cerebral cortex. This circuit, doubtless, is uniquely well developed in the human brain. Pan may speak to Spock and Spock may speak to Pan, but the ideas they can generate, and the way that they generate them, have quite different origins in the brain.

CHAPTER 5

Evolution is just maths

What evolution needs

Darwin's theory of evolution is a theory about biology. It is based on a few core ideas.

First, variation. The members of a species vary in many features. Some plants are taller, some have greener leaves, some taste better to grazing herbivores, some can survive drought while others do better in a bog.

Second, competition. Looking at the teeming world of a small piece of turf, Darwin realised that there is always a 'struggle for existence'. Too many plants are trying to make it on that small piece of ground. Only those that struggle best will survive. Darwin called it 'survival of the fittest'.

Third, transmission between generations. The gene was unknown when Darwin wrote *On the Origin of Species* (1859), but what was clearly known was that many features can be transmitted from parents to their offspring. More successful parents transmit their features to more offspring. Across generations, successful features come to dominate – plants become greener, more able to find sunlight, more unappealing to herbivores.

The basic ideas are just maths. The combination of variation, competition and transmission makes evolution inevitable. Across generations, the probability of successful features must increase.

Very soon after *On the Origin of Species* appeared, it was realised that this is much more than a theory about biology. Across many systems, the combination of variation, competition and transmission means that evolution is inevitable. In his novel *Erewhon* (1872), the Victorian author Samuel Butler applied Darwin's ideas to machines. In the land of Erewhon, machines have long been banned. The argument is that, like animals and plants, machines also evolve, often with harmful consequences – and this time, it is we ourselves who are the agents of variation, competition and transmission. Consider, for example, the way that cars replace horse-drawn carriages as our mode of transport. First, there is variation – we ourselves invent alternative means of transport. Second, there is a struggle for existence – a person who has a car is unlikely to want a horse as well. Third, there is transmission, meaning that every car makes it more likely that there will be more cars. Again, we are our own vehicles of transmission – a car needs a road and a filling station, and once humans have built these, it is all the easier for the next car to be chosen to replace a horse.

The same principles apply to different brands of cars, which compete to be purchased and, with every successful sale, make it easier for the next car of the same model to be manufactured and sold. Or to the environment that cars are used in – with every large parking lot surrounding a new shopping mall, it is easier for the next shop to make a profit at the mall, and harder for the competing shop in the crowded city centre.

Though Darwin's ideas are a cornerstone of biology, in their pure form they are much broader than that. If a system can be described by the basic maths of variation, competition and transmission, the result will be that things evolve, with *On the Origin of Species* being just one particular case.

Darwin himself certainly realised this. In *The Descent of Man*, he points out that words, like animals, compete for their place in the language – a language does not need large numbers of words for the same thing, and the more commonly one word is used and understood, the more it will eliminate competitors. At the same time, I sometimes wonder whether Darwin confused even himself with the language of a 'struggle for existence' and 'survival of the fittest'. 'Struggle' and 'survival' bring to mind familiar forms of conflict, where the weapons are teeth and claws, and survival is a matter of blood and death. When a peacock evolves a large, colourful tail to attract a mate, it can seem that this is not survival of the fittest – the tail can only be an impediment when the peacock must escape a charging predator. Darwin devotes a large part of *The Descent of Man* to the discussion of 'sexual selection' as a special mechanism of evolution – selection by improved ability to attract and breed with a mate. As regards the maths of his theory, however, there is nothing special about sexual selection at all. It is just another case of variation, competition and transmission – just like the case of the bigger jaw, or the popular word, or the cars lined up at the mall.

Ideas evolve too

Most importantly for our own lives, we have the evolution of ideas. While IRMs evolve at the speed of biological evolution, and are passed between generations in the genes, Spock generates new ideas on any topic and at effectively any speed. These ideas too have variation, as we generate many potential ways to pile up stones to make a stable wall, or to dress for a meeting, or to decorate the house for Christmas. Alternative beliefs and practices compete: a person who knows how to build an arch does not try to build a church with stones precariously balanced on top of one another, and a person cannot easily be both a Christian and a Muslim. By many means, finally, ideas are transmitted from one person or generation to the next. We learn from our parents and siblings, we copy what we can see has worked well for somebody else, and our schools are designed to pass on useful ideas. When cultures meet, the practices and beliefs of each tend to cross over to the other. The maths is right for evolution, and like the IRMs of Pan, knowledge and culture also evolve.

Usually, we tend to assume that cultural evolution is a force for good, and that the ideas of today are in some sense an improvement over the ideas of the Victorians or Romans. In one respect, indeed, the evolution of ideas does usually work well. This is when ideas concern the practical reality of the world – how to make a wall stand up straight, or how to think about the rate of change of a variable, or how to avoid a run on the banks. In these cases, the 'fittest' idea has a reasonably simple interpretation – it is the idea that actually works, that allows us to solve problems more effectively and achieve a better outcome. If this is the rule for competition and selection, we get better and better ideas, and achieve better and better outcomes. This is most systematically put into practice in the methods and teaching of science. The products are seen in every aspect of our twenty-first-century world, with enormous volumes of improved ideas and practices for understanding and managing our world.

The situation is rather different, however, for many social

ideas – the ideas, indeed, that perhaps first come to mind when we speak of human 'culture'. As we discussed in Chapter 2, many cultural ideas are riffs on the mental structures already provided by Pan. Many are simple, such as the use of epaulettes to exaggerate the shoulders as a sign of strength, or the dinner guest's gift of flowers and chocolates. Many are much more elaborate, with the military displays, marching children, smiles and firm handshakes when villages or nations come together to trade or negotiate. In later chapters, we will see many more examples in cultural rituals of bonding, aggression, competition for status, loyalty, leadership and more.

In these cases, the rules for what is 'fittest' or most transmissible are perhaps less clear than they are for science but, still, they are not completely arbitrary. Our social lives are built on a bedrock of Pan, and if our cultural practices are shaped heavily by that bedrock, they are likely well adapted to creating smooth social relations.

In some theories of cultural evolution, there is a further, stronger reason for the process to improve the way a society works. This perspective takes the analogy with biological evolution more literally, interpreting 'fittest' not just as most transmissible, but as most socially productive.[1] Here the argument is that whole cultures, like individuals in a species, compete with one another. The successful culture generates more resources, takes control of more territory, perhaps literally produces more offspring. Over generations, the more successful culture invades the less successful, eventually replacing it.

Sometimes the ideas are taken one step further with the proposal of gene-culture coevolution.[2] Here the argument is that culture itself is a change to the environment, and may begin to drive genetic evolution. For example, if a culture promotes harmonious trade, it may succeed and spread. At the same time, the individuals within that culture may be selected for a harmonious disposition. Though genetic selection is doubtless much slower than cultural selection, the idea is that, across a sufficiently long period of time, societies may be improved by these two forces working together.

There is a lot that makes sense about these theories. Obviously, a successful culture does indeed tend to invade those around it. Phenomena from missionaries to genocide show how one culture tries hard to impose its beliefs on others. At the same time, it is tempting to mimic the beliefs and practices of a prosperous culture. McDonald's and Budweiser can spread across the earth, though their actual merits are questionable at best. It seems certain that, through means like these, there must be constraints on cultural evolution, meaning that prosperous, practically successful ideas tend to spread. Later, I shall come back to the likely influence of this process in the spread of religions and moral systems.

All these reasons mean that, very often, cultural evolution leads to 'progress' – in some sense, the later culture works better than the earlier one. We must remember, though, that in the maths of cultural evolution, 'progress' is not the true heart of what makes an idea 'the fittest'. Instead, the fittest idea is just the idea that, for whatever reason, is the most dominant and transmissible.[3] With its high speed, arbitrary content and flexible rules of transmission, cultural evolution is a powerful force on human thought and behaviour – quite possibly, the single most powerful force. But as for where this force will take us, it is just like the Erewhonians and the machines. The evolution of ideas is not easy to predict or to control, and there is no guarantee that it will usually work to make our lives better.

Any ideas can spread

In the world of ideas, evidently, there is a very large volume of material that is entirely arbitrary. These ideas assume dominance not because they work, or because they are built heavily on our instincts, but simply because they are effectively transmitted. Arbitrary though they are, they can assume great power over our lives. Our children would be horrified if Santa Claus appeared through the front door in purple robes. We do not sing 'A Hard Day's Night' at New Year, we sing 'Auld Lang Syne'. These are inconsequential examples, but our world is filled with cases that are not inconsequential at all: for example, when a person emerges from a ruined building waving a white flag, they expect not to be shot.

Many years ago, when I visited Jerusalem, I expected to be beguiled but was just deeply alienated. The steps at the Church of the Holy Sepulchre that for centuries have been the subject of hard-fought conventions over which Christian sect has the right to clean them; the two ends of the Wailing Wall, one for men and one for women; the separate gates on to the Temple Mount that are for Muslims and non-Muslims. I was left with the deepest sense of a city that, for thousands of years, has crystallised all that is most oppressive and at the same time unmotivated in human cultural choices. Every culture in every corner of the world is filled with this arbitrary material, deeply believed but based only on the power of the culture to transmit it.

The maths of an evolutionary system dictate that its behaviour can be quite unstable, easily thrown to extremes. This is especially clear for the case of shared and copied beliefs, competing for dominance in the culture. Let's take any belief X concerning some state of the world. If each person's belief is independent of the others, then combining the beliefs of many people gives a very stable result. Even if any one person has a very low probability of being right about this state of the world, when we take a vote across everybody, with enough people, we are guaranteed to arrive at the truth. As long as the probability of any one person being correct is above zero, when enough people vote, the final probability of the vote giving the right answer approaches certainty. Intuitively we feel that if enough people believe something, surely there must be some truth to it. Simple maths supports this – as long as each person's belief is formed independently of all the others.

In human cultures, however, beliefs are not independent. Instead, weights are linked across brains, as the beliefs and practices of one person influence the beliefs and practices of another. If a priest says that only Muslims should enter through a certain gate, soon many others are promoting the same belief, and similarly, once one army respects the white flag, another is encouraged to follow suit. As one person believes X, so others tend to follow, and the more believers there are, the more the whole system tends to cascade into shared certainty.

Again, the maths of this is very simple. In any system, if the probability of any one element entering the state X is increased by the number of Xs in the surrounding elements, then we have a system that is unstable – it tends simply to run away with itself, ending up with all elements being Xs. It does not matter how the system starts off down the X path – there can be an original good reason for a few Xs, or it can be a complete fluke, something that just happened to tip the system slightly more to X than to some alternative Y.

When it comes to the arbitrary features in human cultures, it is obvious that this happens a lot. Santa Claus did not have to wear red (pre-twentieth-century images depict him wearing green, brown or blue), a flag of truce did not have to be white and a particular section of a wall did not have to be reserved for men.

Our changing world

Let me give two examples of cultural changes that have happened in my lifetime. In both cases, it is arguable whether these changes are for better or for worse but, in both cases, we see important properties of Spock, the ideas he produces, and the factors that promote their transmission.

The first concerns the way professional carers are permitted to interact with children. When I grew up, any adult without hesitation would take a crying child on their knee, give them a hug and let them stay until they cheered up. Sixty years later, any teacher who did this would be taking a severe risk. I well remember how, perhaps 25 years ago, my lovely, gentle sister reacted when she was told that, in her school, she was no longer allowed to hug. Anybody who knows children and who knows Pan is aware that a hug is exactly what a crying child needs. It is possible to do without it but, in making this decision, the adult has thrown away something valuable and effective.

I am not sure why there has been this major cultural swing in the Western world, but it seems very likely that it reflects the mental salience of child abuse. It can certainly happen that, when an adult puts a child on his knee, the reasons are bad and there is major risk of harm. Once you have had this thought, it is almost impossible to decide that the risk is worth it. It does not matter that, for every abused child, there are ten, or a hundred, or a thousand unhappy toddlers who need a hug. (I have no idea of the true figures, but one has to suspect that the true number is very large.) The idea of abuse is simply so salient that it transmits itself as the single most important idea. In my sister's world, the chance to comfort an unhappy child was important all day, every day, but no matter how rare it is, the idea of a plotting child abuser will always trump that daily experience.

This example illustrates all the usual challenges for Spock. The real world is complicated, with many conflicting sides to it. Spock cannot really evaluate the full situation. Exactly how much has been given away by the decision to avoid physical contact? And exactly how much abuse has been saved? Instead, he focuses on a part, and the selected part tends to be salient, dramatic, with a sense of all-importance.

I like my second example because it so vividly shows Spock's tendency to generalise his abstractions. Over my lifetime, in the Western world, attitudes to children have become progressively more gentle. On the night before I started school, my father took me into the barn to learn how to punch corn sacks. He taught me to keep my wrist straight, to put my weight into it, and if the boy was too big, to hit him on the nose. Over the next few years, these skills were used only occasionally, but I was very happy when an angry boy hit me over the head with a makeshift cudgel to know what to do about it. My parents had become adults in the Second World War, and did not like to talk about the friends who had been killed, but without doubt, they thought that when violence came along, it would be better for their children to have some experience of dealing with it.

As regards the relation of parent and child, my mother took rather an ideological line on smacking. She believed, and would explain to us rather definitely, that she regarded all other forms of punishment as unnecessarily extreme. For any minor transgression, she thought it was simply not appropriate to send a child to their room, or to impose some extended burden of disappointment and guilt. She thought a smack let the child know what you thought, that it was over in a second and that after it, everybody could move on knowing just where they stood. We children were perfectly happy with this arrangement. We felt genuinely sorry for those children who had to mooch off to their rooms or who – horror of horrors – might have their sweets withheld.

By the time my own children were at school, a fight in the playground was greeted as a major world event, with parents invited to a discussion with the head teacher and careful consideration of what could be going wrong with these children's relationship. (It always amused me that my son might still arrive home with stud marks from neck to waist after rugby, violence still proving perfectly acceptable as long as it was institutionalised.) As for smacking, in parts of the UK and under many circumstances, this is now illegal – it is regarded, perfectly rationally, as assault.

Now, I discuss these questions with my adult son, and he asks me, 'Do you condone violence?' I reply, of course, that 'violence' is an abstraction, and it is highly misleading to lump all forms of violence together. Human beings are certainly more than capable of this – with the power of abstraction, we can see the resemblance between squashing a mosquito and launching a nuclear bomb – and we can generalise our sympathy too, feeling sorry for the mosquito just as we feel sorry for the victims of Hiroshima. But the real situations are also very different. The question of what treatment allows children to grow up better, happier or more ready for life is simply not the same as the question of whether we should have dropped the atom bomb.

Again, furthermore, the situation is really too complex for Spock to understand in any particularly complete way. It is easy to find evidence for the harms of severe childhood abuse. Not surprisingly, the effects of an occasional smack on the hand will be far more mild, and probably too mild to measure with any certainty. He may not know for sure, but with his yen for abstract ideas, Spock runs away with his broad conclusion that violence is harmful. Culture evolves under the strong influence of ideas that are appealing, simple and general – but usually, faced with complex reality, too appealing, too simple and too general.

As changes like these happen, it is all too tempting to take sides. Our culture wars divide those who like the old ways and those brought up with the new. Both groups believe in their own side, often passionately and angrily. Spock does not like to be reminded of ideas that question his simple view, and there is Pan in here too – as we shall see in the next chapter, the ideas of our own culture give us a vital sense of what we value and who we are. It does not matter that much of this material is uncertain, and much entirely arbitrary. We need to know what we believe, and if necessary, we are ready to fight for it.

PART II

Good and Evil

CHAPTER 6

Us

What we need

In Part I of the book, we have been concerned with the general principles of Pan, Spock and their work together in shaping the human mind. Now it is time to see how these principles illuminate some of our most salient human dilemmas. In Part II, we will begin with a cluster of closely related questions. Love, hate, good, evil, religion, morality and the law.

Surely, just as Viktor Frankl thought, much of the meaning in our lives comes from the people we are close to. Like the innate releasing mechanisms of a bird's bonding ritual, quite isolated events bring us the sense that our lives have importance and worth.

There is sex, adding its aura of glamour to an otherwise humdrum day.

There is community. As an undergraduate, I attended my first scientific talk, and heart beating with mingled anticipation and fear, put up my hand and asked my first ever public question. From the front row, the goddess of our psychology department, Anne Treisman, turned round and gave me what I took to be an approving look, and I walked home across the Oxford parks through the dark winter evening, my heart singing with a sense that I belonged.

There are children. When my first son was born, looking down at my bleeding and smiling wife in the delivery room, I was overwhelmed with a simple sense of happiness and delight, unrelated to the rigours of the preceding few hours. When young friends deliberate about the right time to have children, I tell them that there is simply no point asking themselves. I tell them that, once the child is there, they will be totally different people with totally different values – everything that they care about today (the careers, the restaurants, the holidays) will have been moved away from pole position, as something much much more important takes over their lives. The child fulfils us by just a moment's smile, and we work to achieve the smile just as a bird works to fill the beak of its offspring. Somewhere among our old photographs, I have the most beautiful picture of my wife and small son at his birthday party. To astound him, we have tied up balloons in a net above the table, and as my wife cuts the string, he is looking upwards with astonished delight. When I look at this picture, all my IRMs go off at once.

We are human beings, with our bigger picture, but at the same time, just as Lorenz argued, much of who we are and what we need is just in this one moment.

Animal groups

As Lorenz pointed out in *On Aggression*, animals form groups in different ways. A shoal of fish, for example, is entirely anonymous; there is no sense in which a fish belongs to one shoal and not to another, it is simply attracted to any group of its own species.

Shoaling makes the fish safer from predators, but the only thing that matters is the size of the shoal, not who is in it.

At another level are groups bound by some common cue such as a group odour. To illustrate, Lorenz uses rats, who live peaceably together with members of their own community, but mercilessly destroy an interloper who lacks the group smell. Something similar happens in social insects like ants and wasps, using olfactory cues to distinguish friend from intruder. In these cases, there is a sense of us and them – a rat or an ant cannot simply jump ship to another group, as a shoaling fish could – but the group is not based on bonds between individuals. If a new rat were painted with the correct smell, it would be accepted without disturbance into its new home.

A different kind of group, finally, is based on bonds between individuals. Now, it is a matter of a specific mother and her off-spring, a particular mating couple, a particular tribe whose members are all known and distinguished according to their indi-vidual characteristics and roles. For this kind of bond, we need elaborate internal processes for shaping, maintaining and utilis-ing each individual relationship.

Perhaps all three group types coexist in the human mind. Sometimes, like shoaling fish, we can imagine simply needing others to be around us, perhaps as we are exposed to some immense threat, such as a howling gale or falling bombs. Compared to rats, we are certainly less dependent on smell, but there is still something peculiarly horrifying in Lorenz's descrip-tion of the pack of rats, ruthlessly hunting down and slaughtering the doomed outsider. As I read it, I cannot help wondering whether it horrifies so much because we know that we too are well capable of hunting down an anonymous outsider, not for the wrong smell but for the wrong colour or the wrong prayer book. Most conspicuously, however, human groups are a matter of individualised bonds and relationships. We will help any child in distress, but this is very far indeed from the deep meaning that saturates our feelings for our own, particular children.

Care

Obviously enough, bonding promotes care. In *The Descent of Man*, Darwin proposed that social instincts first evolved from the behaviour patterns of individualised parental care, and this opinion was shared by Eibl-Eibesfeldt. In many mammals and birds, the mother identifies her individual offspring, and only these will be suckled and protected. As Eibl-Eibesfeldt puts it:

> A ewe licks and sniffs her lamb thoroughly after it is born and then will not let any other one push under her. A female sea-lion rubs her muzzle against her baby's and, like the ewe, recognizes it individually soon after it is born and will not tolerate any strange baby near her. Greylag geese attack strange goslings and herring gulls actually kill young that are not their own.[1]

In our own species, we all know the staggeringly strong bond that develops between mothers, fathers, other carers and the individual infants they look after. The human infant is born with a strong tendency to fixate on anything resembling a human face, and to the mother this is intensely rewarding, soon evolving into games of mutual gaze, smiling and speaking in motherese.[2] To the parents, the baby's first smile is a special moment, celebrated and discussed with all the same intensity as the first word or step.

I well remember, when my first son was born, a feeling somewhat resembling stupidity, for perhaps his first 18 months, accompanied his presence in the room. I could nod, chat and smile at the adults in the room but, all the time, it was the little boy on the floor who was the real focus of my attention. I knew that he was a baby like any other, but I felt that he was supremely beautiful, and as survey studies show, this was certainly not unique to me.[3] When my second son was born, this was repeated, and I was absurdly thrilled when a woman standing behind me in a bank queue told me how gorgeous my little boy was.

The bond serves to promote care and, in turn, the behaviour patterns of care promote the bond. In earlier chapters, we saw this in the pheasant offering food to its potential mate, and in the evolution of human kiss-feeding into friendly and sexual kissing. In *Love and Hate*, Eibl-Eibesfeldt gives many further examples.[4] In many species of birds, from terns to tits to ravens, the male and female of a pair offer and exchange food, sometimes with a distinct flavour of the interaction between parent and chick, one member begging for food and the other providing it. In tits, for example, Eibl-Eibesfeldt describes how one partner flutters its wings, resembling the fluttering of a chick, and in response the other partner feeds. In bullfinches, he describes ritualisation into a gesture without food, a simple 'flirting with the beak'. The friendly pawing of a dog, he speculates, is ritualised from the pawing movement used by a puppy to release milk from the mother. He describes how, pursuing a female, male hamsters imitate the cry of a young hamster that finds itself outside the nest.

The gift used in bonding rituals need not be food. As we have already seen, for example, it may be nesting material, which the partner may seize and build into the nest. Also very common – and as we shall see, especially important in primates – is social grooming. Just as mothers clean the young, many birds and mammals nibble at neck feathers, remove parasites, scratch and stroke. We love to scratch or stroke horses, dogs and cats, and, in turn, they love to be scratched or stroked, often throwing themselves to the ground and rolling over to invite this behaviour. Certainly for one partner, and it appears for both, this grooming produces a rush of individual affection.

Against the outsider: The triumph ceremony

In many animals, mating pairs or larger groups cooperate to defend their territory. This can be a pair of fish staking claim to their share of an aquarium, or a pair of geese driving off others from their family, or a hyena clan warring for control of a rich hunting area. Lorenz pointed out that an aggressive, territory-defending species has an evolutionary problem. Aggression must be inhibited within the mating pair or the larger group – all these individuals must be able to inhabit the same territory. But aggression must be released by other, similar individuals who appear in the same place, sometimes sporting much the same releasing signals. Lorenz saw individualised bonding as a solution to this evolutionary conundrum.

To support this idea, Lorenz drew on the ritualised displays used to promote individual bonding. Using examples from reef fish to cranes on their nest, he described bonding displays that have clearly evolved from the usual aggressive displays directed to an outsider. But these displays are changed in one critical respect. The male cichlid may be charging his partner, apparently on the verge of attack, but at the last moment, he swerves to one side and directs the display past her, perhaps at a hapless neighbour.[5] In the dance of the cranes:

A bird rears up before another one and unfolds its mighty wings, its beak pointing towards the other bird, its eyes fixed piercingly on him, the very image of ominous threatening; so far the appeasement gesture resembles the preparation for attack; but the next moment the bird turns this exhibition of his own frightfulness away from his opponent by a right-about turn; and now, still with widely-spread wings, he presents to his partner his defenceless occiput, which, in the European crane and many other species, is decorated with a little ruby red cap . . . Now the crane turns again towards his friend and repeats this demonstration of his size and strength, swiftly turning round once more and performing emphatically a fake attack on any substitute object, preferably a near-by crane which is not a friend, or even on a harmless goose or on a piece of wood . . . The whole procedure says as clearly as human words, 'I am big and threatening, but not towards you – towards the other, the other, the other.'[6]

Or from Eibl-Eibesfeldt:

A male rhesus monkey that wishes to make friends with another begins by remaining near him. He tries to win favour by offering to delouse him. In addition to this he will make sudden attacks on passing monkeys; in other words he tries to draw his prospective friend into a joint action, and sometimes succeeds. Once the two of them have beaten up someone together they are good friends.[7]

For Lorenz, this process of bonding by redirected aggression reaches its apogee in the 'triumph ceremony' of his beloved greylag geese.[8] The triumph ceremony, he thought, lies at the heart of the entire social life of the greylag. It underlies the bond between mating pairs, families, individual pairs of friends and whole flocks. Lorenz proposed, very seriously, that it is a precise analogue to human love. Of course, he did not mean that the emotions of the bird feel like our own emotions – this is something we could never know (just in the same way that we never really know what the emotions of another person feel like on the inside, though we reasonably assume that they feel like our own). But he thought that, by convergent evolution, the love between greylag geese had evolved with astonishing details to resemble the love between human beings, all cemented together by the mechanism of the triumph ceremony.

As described by Lorenz, the triumph ceremony has two phases. In the first phase, called rolling, the bird rushes away from the partner or family, head forward and up, making a furious trumpeting attack on some real or imagined opponent. Having completed the attack, preferably victorious, the bird enters the second phase of the ceremony. It returns, neck stretched low and again with an impression of attack, but now subtly directed off to one side. Now there is a new cry, called cackling, and partners or the whole family group may cackle this greeting together, heads held close together. Lorenz describes how, throughout the activities of the goose group, there is always an undercurrent of this friendly cackling, rising to passionate intensity in the full triumph ceremony. He calls it a 'passion of togetherness'.

Greylags pair for many years, perhaps sometimes for life. When a young gander 'falls in love', he offers the triumph ceremony to the goose of his choice and, if he is lucky, she is interested and will accept it. This usually leads on to mating, but Lorenz is quite clear that the triumph ceremony, not copulation, underlies love in the greylag. Geese bound by the triumph ceremony nest together, walk together and cackle together; it may well be that a gander has another sexual partner, one who would cackle with him if he would, but instead he finds some out of the way place to mate with her, then hurries back to cackle with the family.

A young gander who has fallen in love, and is offering the triumph ceremony, undergoes a complete change of personality; he stands proud and erect, takes wing without effort (Lorenz says that flight comes 'as easily . . . as to a hummingbird'), and enjoys racing back towards the object of his attentions and landing with a flurry in front of her. She is more shy; if she is interested, she watches him out of the corner of her eye, but only gradually and gently joins in with the cackling. Sometimes the young female in love appears starstruck; she does not initiate the triumph ceremony yet, but simply turns up where she expects the gander to be.

When a pair is broken, perhaps by the death of one partner, new triumph ceremony partnerships can be formed, but often these are not like the first love. Now the gander is more likely to copulate outside the family. Sometimes it is two males, not a male and a female, who form the triumph ceremony bond, and sometimes, with great difficulty, a female then manages to enter the group too, and form a ménage à trois. Sometimes there is jealousy, with ferocious exchanges between two geese trying to share the ceremony with the same partner. There is grief, with an increasingly desperate and far-ranging search for a partner who has been killed, and then a descent into fearfulness and low rank. Reading Lorenz on the triumph ceremony is like reading a Mills & Boon.

In Chapter 1, we saw how IRMs have a need to be discharged, building up over time if no suitable opportunity arises. For the greylag, there is a deep need to be near the partner, for the triumph ceremony to be discharged and the bond to be reaffirmed. There is a bold idea in here. Lorenz did not think that the bond exists separately from the need for reaffirmation.

He thought that the need to perform the triumph ceremony *is* the bond. On this view, the bond is not abstract. It has specific content – a craving to perform specific behaviour.

At first, this seems odd, but I think it is not hard to see something similar in ourselves. A man has lost his wife. He may say, 'I miss her,' but there is much more than this. Surely it moves us more when he says, 'I miss laughing with her,' 'I miss cooking with her,' 'I miss going to the movies with her.'

In geese and other animals, Lorenz thought that bonds produce what he called 'home valency'. The presence of the partner brings a feeling of safety not unlike the security of the home nest; together, the partners are ready to challenge the world, in the same way that they will drive off an intruder from the nest; left alone, they are weak, like an animal far from home. Again, this is like a Mills & Boon. We know we are this way.

Many years ago, a group of friends took me to Laser Quest, and I had enormous fun dodging through the mazes of the dark interior, trying to be the last one to be shot. The next morning, I awoke to the realisation that, in the night, I had been constructing a strategy for the next time I went. My wife and I would go together; we would dodge through the tunnels standing back to back; this way nobody could take us by surprise and we should be unbeatable. In my half-sleep, I had been concocting a literal version of the expression 'I've got your back.'

The human group

Now from the goose to us. Much modern research supports Viktor Frankl's thought that a large element of meaning in our lives comes from the people around us. Both psychologically and physically, friends are important – people with friends are happier, better protected against life stress, better at recovering after surgery.[9]

In our modern civilised world, we collaborate on a grand scale. With language, an understanding of one another's desires and intentions, and the ability to pass large bodies of knowledge between generations, we are able to cooperate in massive group endeavours from understanding and splitting the atom to exploring the limits of the universe. These collaborations with hundreds to millions of other people, however, are not the same as friendships. If we want to understand the Pan of our own friendships, we should look to the environment in which they evolved. Up to around 10,000 years ago, before the arrival of agriculture and civilisation, humans evolved in hunter-gatherer groups, extending from a few families to an upper limit of around 150. Human beings have many social instincts, forming the bonds and managing the relationships we need for a society of that sort. Beyond that, as we shall see in later chapters, we need something different – not the power of friendship, but the power of Spock.

This theme has been developed at length in the work of the psychologist Robin Dunbar. Dunbar proposes that, like an onion, our relationships are organised in layers, with progressively reducing intimacy and contact as distance from the centre increases.[10] At the first level, which Dunbar describes as 1.5 people, are oneself and perhaps one's partner. (In my early twenties, when I first started to have married friends, I was struck by a young Australian woman who said, 'You know it's real when you begin to find you're saying we instead of I.') At the next level, there are around five intimates, people we would go to in times of severe distress. So the layers expand through around 15 and around 50 to an outer layer of around 150, the rough upper limit on the size of a hunter-gatherer community, and the upper limit on the number of people we can say we have a meaningful individual relationship with. Of course, we can name and recognise more people than this, but now they are just people we know of, not members of a shared community with its separate, individual relationships.

Dunbar supports these ideas with examination of social media groups, face-to-face contact times, historical village sizes and more. We may be uncertain about the exact division into layers, but there seems little doubt over the broad principle of levels of intimacy and the rough upper limit of 150 individual personal relationships. Sometimes this upper limit is called 'Dunbar's number'.

Much of this structure is based on kinship. At least the layers of 5 and 15 are often filled, wholly or entirely, by members of the extended family. In terms of dependence, reliance, emotional involvement and willingness to support, family members have a highly privileged status, just as they do in greylag geese or any other animal living in related groups. When we wish to express a bond, we talk of brothers, sisters, a motherland and a fatherland. The degree of genetic kinship decreases as we move out across the layers, though in a traditional, enclosed community of around 150 people, it is likely that everybody will be related to some degree.

The structure is based in part on commitment of time. Dunbar estimates that around 40 per cent of our social time is devoted to the inner circle of 5 people, with about 20 per cent given to the remainder of the top 15. If the time allocated to a person is much reduced, the relationship rapidly loses its intimacy. The importance of time is interestingly different between men and women; for women, the time spent talking is very important, but for men, it is the time spent doing things together, even if all they are doing is having a pint.

This need for an investment of time means that the layers are self-correcting. If a new person moves into the closest layer, the time left for others diminishes. This is most apparent when a person falls newly in love; commonly, at this point, other close friends lose time and intimacy. When we move to a new place, we rapidly fill the layers with new friends until we have the right number. People with large extended families have fewer non-family members in their inner, most intimate layers; the family members take up the available time and the available slots, and others have to make do with what is left.

When we search for the human version of the triumph ceremony, then, what we are looking for is a set of bonding mechanisms that can create a group of around 150 people. Within this group there are varying levels of intimacy but, somehow, all these people have to think of themselves as 'us'.

In our primate relatives, bonding depends heavily on social grooming. Monkeys and apes spend much time grooming one another and, of course, though we generally will not need parasites to be removed, we also promote intimacy by stroking, caressing and cuddling. In our case, however, this mechanism does not really extend up to a wide friendship group. It depends a bit on the individual culture but, by and large, we do not take close caressing far beyond the most intimate layers of our social network.

In some of his early work, Dunbar charted the time spent grooming across different primate species.[11] The idea was that, if grooming promotes the individual bond, then the total time an animal needs to spend will increase with the size of the group. Indeed, this happens, but the time spent reaches a maximum at around 15–20 per cent of total daytime hours (of course, no animal can spend 100 per cent of the day grooming its friends), and at this point, the group size is still only around 50. If *Homo sapiens* is going to have a group size of 150, something else is going to be needed.

Cichlid pairs have the aggressive display directed at a place beyond the partner. Greylag geese have the complexities of the triumph ceremony. In our case, evolution has come up with a range of parallel solutions. Different though they are on the surface, they all promote that gushing sense that we are 'us' – that we are in this together. Some have an evident basis in the behaviour patterns of other primates, while some are specifically human. Together they form the IRMs and the Pan of social cohesion.

In the Notes, I have added a link to Playing for Change.[12] Take a break from my book and have a look. Since my brother first directed me to this group of musicians, I have been saying that all world leaders, on the day they take office, should be strapped to their chair and forced to watch them. In Playing for Change, we have all the power of human bonding mechanisms showing that, no matter who we are, we are all one.

Music and movement

First, there is the music itself. One summer, when I was spending a few weeks on the farm with my mother, I decided to introduce her to Bruce Springsteen. I put on a CD in the kitchen, and together we sat at the table listening. We had spent a few close weeks together in the house, and as Bruce played 'American Skin (41 Shots)', I was suddenly, unexpectedly overcome by an extraordinary welling sense of importance and togetherness.

This, I imagine, is the evolutionary root cause for the existence of rock festivals. Through the rain, the mud, the disgusting toilets and the general discomfort, around 200,000 people track to Glastonbury every year for this sense of togetherness.

The brilliance of Playing for Change is that musicians from the whole world come together to create their version of rock classics. These people are different, everything about their lives and their experiences is different, but with 'Gimme Shelter' they are one. It is perhaps an exaggeration to say that music can save the world, but it is an exaggeration based on something fundamental about humanity.

Of course, this does not have to be rock. In opera duets, we have the alternation between the two soloists, gradually merging into a passion of simultaneity as the duet reaches its climax; much the same thing is seen in mothers and infants playing vocal games together, with alternation merging into simultaneity as excitement builds.[13] Those of us who know nothing much about it perhaps think of a symphony orchestra as a rather intellectual institution, but it is not. I asked my daughter-in-law, who plays music I cannot begin to appreciate on the bassoon, and she said: 'I get [the feeling] in any group that's larger than a trio, I think, although the feeling is much more intense in a full symphony orchestra. I asked my colleagues and one, who is a cellist, says it's particularly strong in a string section because often the aim is to sound like one instrument, with no voice more noticeable than the others. This happens in a less obvious way between sections across the whole orchestra. I think it's partly the act of creating something together, but also the beauty of what we are creating – it feels so special to be a part of it.'

Rhythm, music and dance are human universals, found across cultures and used in all kinds of settings, especially rituals of bonding and sharing.[14] What is shared may be a memory, a celebration, a set of religious ideas, a plan for joint action, such as collecting the harvest or waging war. Often there is drumming and shared swaying to the beat.

There is really nothing like this in even our closest relatives such as gorillas and chimpanzees. Sometimes these animals may show a kind of drumming,[15] perhaps on some resonant object, but nothing like the extended, synchronised, highly social drumming of the human being. In our brains, there is a strong response to 'the beat' that is not seen in other primates.[16]

Why should shared rhythm be so important in our species? As we saw in Chapter 2, humans are special in the depth of their shared action, from two people manipulating a bed up a set of stairs to the immense rhythmic structure of the symphony orchestra. For things that we do together all day every day, success depends on simultaneity or more complex temporal coordination. We use rhythms to control this – 'one, two, three, PULL' – and sometimes this is explicitly musical (think of hundreds of sailors singing, 'Heave, ho, my hearties . . . '). After the manner of rit-ualisation throughout animal behaviour, the thought would be that we evolved rhythm to coordinate our behaviour together; and then rhythm itself became incorporated into powerful rituals cementing the sense that we are one.

Armies, of course, have perfected this use of simultane-ity to the highest degree. One of the best examples I know is Leni Riefenstahl's staggering *Triumph des Willens* (*Triumph of the Will*), the Nazi propaganda film based around the 1934 Nurem-berg rally. The film is Pan writ large: the shadow of the Führer's plane, descending to a waiting world from the sky, the power of Hitler's oratory that can be felt without understanding a word of German, and the steadfast, synchronised, machine-like marching of the immense bodies of troops. It is perhaps no coincidence that Leni Riefenstahl had begun life as a dancer. From the other side of the Second World War, here is a description of William McNeill's American military experience:

> Words are inadequate to describe the emotion aroused by the prolonged movement in unison that drilling involved. A sense of pervasive well-being is what I recall; more specifically, a strange sense of personal enlargement; a sort of swelling out, becoming bigger than life, thanks to participation in collective ritual.[17]

'A sense of pervasive well-being' is not a bad description of the high that tens of thousands of people felt swaying to Queen at *Live Aid* in 1985.

By sheer luck, I was in the crowd in 2011 when, 30 years after Pink Floyd's acrimonious split, and with only vague rumours that something might happen, David Gilmour reappeared for one night in Roger Waters's recreation of *The Wall* tour. As Gilmour's tiny spotlit figure appeared on top of the wall and began to sing 'Comfortably Numb', my son leaned over and yelled in my ear 'David Fucking Gilmour!!' and around the vast O_2 arena grown men were weeping at the sight of the bond (however temporarily) reformed.

Robin Dunbar proposes that, at least in part, this swelling, euphoric sense of togetherness reflects the release of endorphins into the brain. When people sing, dance or row together, their sense of togetherness increases, as shown by direct comparison of before and after ratings, and at the same time their pain threshold is increased, a potential indication of endorphin release. Some of this sense of togetherness can be offset if, before the experience, the person is given a drug that blocks the endorphin effect. Endorphins are internally released opiates; on Dunbar's account, these internal opiates have been coopted in evolution to produce the high of togetherness.[18]

Laughing together

Music is not the only way to get the high. We all know the help-less, bubbling sense of kinship that comes from a long evening of laughing together with our family or friends. Music and laughter have quite different evolutionary backgrounds, but they bond us together in much the same way.

Laughter and its precursors are quite similar in humans and chimpanzees. In Chapter 2, we introduced the 'play-face', with mouth excitedly open and teeth bared, used by small children from many cultures to signal a friendly desire to interact, and strongly resembling the play-face of tussling chimpanzees. The play-faces of humans and chimpanzees, indeed, are so similar that each species recognises the play-face of the other.[19] As the intensity of the interaction gathers force, the play-face of the human toddler often leads on to outright laughter, and chimpanzees also laugh outright when playing together.

In the exposed teeth of laughter, it is not difficult to recognise a ritualised gesture of aggression, very much as we see in the triumph ceremony of the greylag goose. Eibl-Eibesfeldt linked laughter to mobbing; in many social animals, the group comes together to drive off a common enemy and, in monkeys and apes, this mobbing includes baring the teeth and uttering rhythmic threat sounds. In our case, there is an obvious element of threat in the eliciting conditions for laughter. We may give a grunt of disgusted laughter when we bump into a table or drop our lunch, and in jokes, there is often a strong element of shock as the denouement is quite different from what we were led to expect:

'What do you call that useless bit of flesh at the end of a penis?'
'A man.'

In a later chapter, we will come back to the unabashed aggression of laughing *at*.

We do laugh at jokes, and when people are asked what makes them laugh, they will usually focus on humour. When people actually keep a record of their laughter, however, it turns out that humour is not the main point. The main point is that we laugh when we are together with friends – around 30 times more likely than when we are alone.[20]

In humans – much more than in chimpanzees – laughter is extremely contagious; the sound of another person laughing can make it almost impossible to resist joining in. This laughter contagion is much stronger between friends than between strangers, and leads to a sense of uncontrollable mutual glee. My friend Sophie Scott analyses a famous episode from a pair of radio cricket commentators, Brian Johnston and Jonathan Agnew, slowly failing to maintain their professionalism as they each begin to laugh on air at a (very minor) joke.[21] This clip reportedly caused drivers to pull to the side of the road as they began laughing too; its immortality was reflected in a radio programme dedicated to its memory, made by the BBC 30 years later (*Test Match Special*'s 'The "Leg Over" 30 Years On'). Sophie's own TED Talk is well worth watching for both the science and your own experience of laughter contagion.[22]

Brain imaging studies show that laughter is controlled in part through the periaqueductal grey.[23] This is an ancient brainstem nucleus whose role in instinctive movement patterns I discussed in Chapter 4. Like music and dance, laughter leads to increased pain threshold, suggestive of endorphin release.[24] Usually, when we laugh, we laugh together, and with this ritualised aggression, we bind ourselves together against the threats, false steps and perplexities of this crazy world.

Working together

Dancing together has a strong, specific effect on our social bonds, but it doesn't have to be dancing. From Quaker barn raisings to sports teams to the family, we all know the sense of belonging that comes from working together. As Dunbar notes, just spending time together is important, but the sense of belonging grows stronger with work on a shared enterprise, with sustained and coordinated effort, with difficulties and challenges faced together, with the importance of the result.

Quoting from an early writer, Wikipedia describes barn raisings as 'occasions of community good-feeling, solidarity and festivity . . . Customarily the women of the families involved prepared hearty lunches for the builders and completion was celebrated together with a feast and dance – often till dawn.'[25] My own childhood on the farm was filled with this kind of solidarity and, 60 years later, it remains deeply precious to me.

For example, every August, there was the corn harvest. The event is critical – in England the corn (for us, a mixture of barley and oats) ripens just before the autumn turns rainy, and every year, tension rose with the hope of a few dry days for the corn to be harvested and the straw to be brought in. When those days came, the whole family watched anxiously for the combine harvester to arrive. In my early childhood, the corn was still collected in sacks from the field; late into the summer evening, until it was too dark to work, the men arrived with sacks piled on a trailer, and waiting in the barn we children would cut the string on each sack, roll back the top and organise everything in rows with the corn open to the air. Later, when the straw bales were collected, we were on the trailer in the field, packing the bales layer on layer as the men passed them up on pitchforks. As we grew older, our roles changed; now sacks were a thing of the past and we drove the tractor along beside the combine so that the grain could be shot into the trailer. My grandmother brought drinks and food; when it was dark we were tucked wearily into bed.

Or again – when I was about seven or eight, my father decided to replace the concrete of the cow yard. The old concrete had to be broken up and, one evening, while my father was milking, I decided to have a go myself. I got the smallest sledgehammer and began smashing at the edge of the stretch to be removed. At the end of milking, my father appeared and was amazed. (I expect he was not as amazed as he seemed – at seven or eight, I can't have achieved much.) Now, at the age of 70, I still love the sight and feel of a sledgehammer.

In *Baby and Child Care*, Benjamin Spock advises on teaching children to contribute to home duties. He says: 'It . . . helps a lot to assign children tasks that they can do in the company of other members of the family, whether it's dish-drying or lawn-mowing. Then the grown-upness of the task and the fun of helping spur them on.'[26]

This same thing, surely, is at the heart of a successful marriage. Of course, my wife and I have shared the joys of sex, laughter and holidays, but over 45 years, there is much more than this. When we were first married, we were completely broke; at the end of each month, we pored together over cook-books thinking what we could make with cabbage. When we first papered a ceiling together, we had not fully understood the properties of the whitewash we were papering over; when the ceiling was finished, we went out into the sunny garden to celebrate with a beer, and when we came back in, the paper we had just glued up was hanging down again in streamers. (Luckily, it had brought the whitewash down with it, so all we needed to do was paper it again.) For two decades we conspired to manage, feed and delight the children as they grew from screaming infants to thoughtless teenagers. We expect and hope to face the challenges of old age together; we are both very moved when Alison Krauss, delivering her version of 'When You Say Nothing at All', sings of the unspoken union of shared and unquestioning support.[27]

After all this, of course we are bonded! I will come back to more experimental evidence in a later chapter, but really, this is something that everybody knows.

Gifts

Earlier we saw how gifts of food, nesting materials and so on are used by many animals to allow approach, cooperation and often mating. Across many cultures, humans also promote friendliness with giving of all kinds. From our own experience, we know well how babies play give-and-take games with their parents and siblings, especially with food but also with many other objects. Giving and receiving gifts is accompanied by smiles and exclamations of delight.

In *Human Ethology*, exactly similar sequences appear in Eibl-Eibesfeldt's films of many cultures, with all the happy generosity of a Yanomami infant feeding its older sister, and all the conflict of a Biami boy who wants both to give and to keep. The adult versions are more culturally ritualised, as we hand over whisky to a Japanese host or flowers to a visiting princess, but still with many of the same, immediately recognisable features.

Once again, there can be striking similarities with sharing in our primate relatives. Jane Goodall describes how chimpanzees extend their hands to beg for food, and we understand the gesture without effort because it is our own.[28] When a chimpanzee band has captured some valuable prey, the owner shares it out, tearing it into pieces and handing it out while the others sit with hands raised.[29]

The exact same sharing of meat is seen in many hunter-gatherer cultures, with correct sharing bringing social status. Like many other animals, we make especial use of food exchanges. Eibl-Eibesfeldt describes how, when a stranger visits a Maasai kraal, he waits outside until relations are opened by an old woman bringing milk.[30] Similarly, a visitor to another tribe in the Walbiri of central Australia brings a gift perhaps of kangaroo meat.

We invite family, friends and associates back and forth for brunch and dinner parties. The sense of a bond may be promoted not only by giving and receiving, but simply by eating together. Another example from Eibl-Eibesfeldt: in the Patasiva of the Moluccan Islands, a wedding is not sealed until all have eaten papeda together.[31] In our case, we have the ceremonial wedding cake.

As anthropologists have pointed out since Bronisław Malinowski in the 1920s, exchanged items need have no practical value.[32] Reciprocal exchange in itself functions to cement social relations. Malinowski's famous example is the Kula exchange system of the Trobriand Islanders of Papua New Guinea. In the Kula system, as described by Malinowski, gift objects were constantly in circulation between communities of the different islands, including a mother-of-pearl necklace circulating clockwise, and mother-of-pearl armbands circulating counterclockwise. At each exchange, men from one island might cross dangerous waters to deliver the gift to the next, but these gifts were important only for their symbolic, not their practical value. Once received, the necklace or armband would be retained only for a while, and was then sent off again around the exchange circle. The Kula system was without any commercial point; instead, it served to cement relations between potentially hostile neighbours.

Across cultures, the gifts that are exchanged in greeting and bonding can be arrowheads, tools, wives or much more. In our own Western culture, we have Christmas, and once again, the commercial value of the gift is really beside the point. In extreme (in my opinion, rather degenerate) cases, each person simply gives money to the other, an objectively pointless activity that nevertheless fulfils the function of social exchange.

Gifts or other practical favours can be used to defuse hostility, and as Benjamin Franklin famously showed, it is the giver as well as the receiver who feels the effect. In the 1730s, when experiencing ongoing conflict with a colleague on the Pennsylvania legislature, Franklin settled on an unusual course of action. As his rival had a great library, Franklin asked to borrow a book, returning it in a few days 'expressing strongly my sense of the favour'. Having done one favour, the erstwhile rival was ready to do many more, and as Franklin said, 'Our friendship continued to his death.'

As we all know, and as anthropologists have observed in many cultures, gifts must be made with care. It must be possible for a gift given in one direction to be reciprocated with a balanced gift in the other; a gift too large to be reciprocated takes on an aggressive aspect, with the receiver then beholden to the giver. In the potlatch festivals of Native Americans in the Pacific Northwest, chieftains strove to outdo one another with feasts of extreme generosity, sometimes extending to pointless destruction of their own property. Eibl-Eibesfeldt quotes the lines of a potlatch host's song:

I am the great chief who makes people ashamed . . .
I am the first to give you property, tribes . . .

Bring your counter of property, tribes, that he may try in
vain to count the property that is to be given away by the
great copper maker, the chief . . . [33]

Shared beliefs and experience

Music, laughter, shared activities and gifts are all quite concrete.
Human bonds, however, are based just as importantly on shared
ideas, beliefs and habits. Among critical factors determining
friendship, Dunbar lists shared dialect, place of growing up, interests, tastes, worldview – 'the set of beliefs and rituals that remind
us who we are, where we come from and why we form a single
community with a common set of values and convictions'.[34] The
more of these things two people share, the more intimate is the
friendship likely to be.

In *On Aggression*, Lorenz also emphasised the importance of
shared beliefs and habits, along with the ceremonies and rituals
we use to support them.[35] At Christmas, we repeat the ceremonies
of decorating the tree, bringing in holly, inviting the same friends
for the same eggnog party, waking on Christmas morning for the
awed excitement of the children's stockings and the same Christmas breakfast. Across cultures, religious services reinforce a fixed
set of beliefs, very variable in their content, but with the same
critical function of cementing the group's sense of themselves. In
many cultures, initiation rites attend the passage on from childhood, with instruction in the practices, roles and beliefs expected
of an adult. A major component of ceremonies and rites is the
use of shared stories, perhaps stories of ancestors, or the gods,
or the creation of the world, all reinforcing this culture's body of
common knowledge.

These are the inventions of human culture, but Lorenz saw

parallels with animal behaviour and the IRM. Like instinctive behaviour patterns, he suggested that cultural habits have their own need to be performed, and their own satisfaction in being completed. Perhaps this happens because Pan has provided the outline need for knowledge and habits to be shared; Spock has simply attached specific content to this need. We do not just carry out the rituals of Christmas – we love them, with a sense that once again our community is reaffirmed.

In birds such as chickens and geese, Lorenz described the phenomenon of imprinting: the chick is born without knowledge of a parent bird's appearance, but in a critical period just after birth, it seizes on the first roughly suitable moving object it can find, and thereafter follows it as though it were the parent. Lorenz thought that something similar happens, towards the end of puberty, in the formation of human beliefs and values. At this period, he thought, we learn what sort of person we are, how we act and what we stand for, and for the rest of our lives, our thoughts and actions are strongly coloured by this image.

Backing up the shared identity of our beliefs and aspirations are many arbitrary style choices, from the colours of a football scarf to the headband decorations of a !Kung Bushman. We know who we are and we advertise who we are. We are 'us'.

And more . . .

We could go on. Since Freud, it has been suggested that bonds can be formed by shared identification with a charismatic leader.[36] It is easy to imagine that, besides eating together, other activities of hunter-gatherer societies would have acquired their own specific bonding function – perhaps shared childcare, shared hunting, shared cooking. It sometimes occurs to me, as I sit for an hour staring into a fire (especially a campfire), that perhaps this love of staring into the fire together is also genetically programmed. Certainly it is easy to imagine that, in the night of the African savannah, there would be little future for a child who did not love to stay with the group and stare into the fire. (Sometimes I go further and wonder whether much of the appeal of television has the same basis; very often, there is little merit in the actual content of the moving images, and for me at least, a falling log is infinitely more absorbing than another episode of *Friends*.)

In tune with Viktor Frankl, we might also consider the bonding role of shared suffering and grief. In a meeting between Yanomami villages, Eibl-Eibesfeldt describes how, late into the evening, women of the two communities gathered to weep over their dead.[37] Humans are the only species to use tears not just to keep their eyes moist, but to express deep emotion. The welling sense of 'us' produced by laughing together is almost eerily similar to the same sense when we have shared 'a good cry'.

There is also something big and obvious that I have not come to yet. As Lorenz argued, in many species, the bonding of 'us' is strongly based on shared battle against 'them', and it is no surprise that, in a heavily social species such as *Homo sapiens*, few things bond more reliably than struggle against a common enemy. For now, though, I am putting this one on hold. In Chapter 8, we come back to the two sides of 'us' and 'them', with the bonds between 'us' excluding 'them', and the threat of 'them' strengthening the bond of 'us'.

Some of what we have seen in this chapter is fairly pure Pan, like the infectious, uncontrollable laughter of Johnston and Agnew taken over by their periaqueductal grey. Some show the interaction of Pan and Spock, with Pan providing an instinctive framework and Spock filling in cultural details. While the need to share rituals and beliefs is likely an innate human universal, it must be Spock who dictates that Santa will put presents in a stocking, or that a particular hut will be used for the initiation ceremony. Pan provides broad themes and structure, and Spock fills in details, just as learning adds the content of a particular language to an innate blueprint for communication.

With all of these forces acting together, we build bonds to partners, family, friends, community and, as Viktor Frankl proposed, living out these relationships is central to our sense of meaning and purpose. We know the joys of preparing and eating food together, of seeing a child's delight as it unwraps a gift under the Christmas tree, of relaxing with a beer after we have helped the neighbour to repair his fence. Like the greylag goose discharging its need for the triumph ceremony, we have an enduring need to laugh together, to plan together, to meet challenges together . . . and when we do these things, our lives feel deeply right.

CHAPTER 7

Keeping things together (1)

Human dealings

Forming a group is only the start. Much more is needed to manage the countless social relationships within this group, to build a complex structure of roles and interactions that make it work. Amid the rich variety of the world's cultures, anthropologists and cross-cultural psychologists discern broad patterns of social interaction.[1] The stability of these broad patterns across cultures strongly suggests an innate basis. Again, it is like language, with a broad, likely innate structure forming the framework on to which countless cultural details are added. There are many different moralities, dictating how we should conduct our lives and how we should treat others, but under these varying moralities are strong common themes.

Our lives are built around countless cultural details, and as Chinua Achebe tells us in his 1958 novel, when cultures collide, *Things Fall Apart*. In this chapter, I deal with the other side of this story – the idea that, across cultures, innate structures of social interaction are there to keep things together. Without these elaborate, powerful constraints on how we conduct ourselves, there could be no society and nothing to fall apart.

For this chapter, I shall lean heavily on the work of the anthropologist Alan Page Fiske and, in particular, a paper published in 1992 in the *Psychological Review*.[2] The paper gives a jaw-dropping sense of an author who had read and thought about everything that could possibly bear on human society – Sigmund Freud, Karl Marx, Émile Durkheim, Maragaret Mead and more, and more – but it was Fiske's own experience of different cultures, and, in particular, his fieldwork among the Moose of Burkina Faso, that most strongly shaped his theory.

Fiske describes four core modes of human interaction. As soon as they are described, they are intimately familiar. It is Eibl-Eibesfeldt again, but now at a more abstract level, with broad structures of social interaction rather than the local details of a smile, a kiss or a threat.

Communal sharing

Fiske calls the first mode *communal sharing*. In communal sharing, we are one. There is no counting, no equal division of the food or the land – like the parts of one body, each person simply takes according to their needs. There is no keeping track of who puts in what – people 'just pitch in and work until the job is done, treating it as a joint responsibility, whether the task is painting the church, defending the citadel, or digging out survivors of a building collapse.'[3]

Singing 'Imagine', John Lennon invited the world to dream of a future with no ownership, no struggle for possession, all our goods shared.[4] On the radio, I once heard a cheerful commentor say that, this time, John Lennon had really got it wrong, and in our Western world, it is hard to bring off this particular act of imagination. In much of the world, however, it is the norm. According to Fiske, communal sharing is the dominant principle in many hunter-gatherer communities: 'people share the meat of game animals across the whole band: the hunter who killed the animal often ends up with less than many others, and people share food, tools, and utensils with anyone who asks for them.'[5] It is 'the dominant, ideologically preferred firm of group organization in the stateless, food-harvesting societies of . . . Southeast Asia, Australia and Oceania'.[6] Above all, of course, it is conspicuous within the family. We pour effort and resources into our children without a thought that the debt could never be repaid. It is not a debt. Our children are part of ourselves. Rather winningly, Fiske points out that we speak of others in our community as 'our kind' – and 'kind', of course, has another meaning.

Communal sharing is not only a matter of cooperative labour, support and division of resources. In this mode, the group makes decisions by searching for 'a joint judgment that transcends the separate attitudes of the participants'.[7] Perhaps most important, the members of such a community derive their beliefs, their actions, their sense of self from the group around them. There are shared meals, rituals, religious beliefs and a strong sense of contamination when somebody does not conform. Elsewhere in social psychology, this is picked up in 'social identity theory'.[8] The sense of the self depends strongly on membership in the group. Often, the group is bound together by a sense of how things have always been done, perhaps as laid down by common ancestors. Fiske says:

> Among Moose, for example, men who are all descended from a common male ancestor through the male line tend to live together, pool their labor, eat together, make religious sacrifices together, help each other to court wives, and feel that they share a collective fate – they are jointly at risk when any member violates a taboo.[9]

Communal sharing is in many ways a dream. It is what communism could be, if only there were nothing else in the human mind. Of course, there is.

Authority ranking

In communal sharing, everybody is equal. In the next mode, *authority ranking*, there is a linear ordering. Those higher up in the rank get a greater share. They take precedence in where they sit, in the size of their dwelling, in the wealth of their possessions. Indeed, conspicuous consumption may itself be used as a sign of rank. Often, those higher in rank will contribute less to manual, menial or unpleasant tasks. Those lower in rank may be expected to pay tribute and, often, superiors may be expected to take whatever they need from inferiors, with no expectation of return or repayment.

At the same time, higher rank brings responsibilities and expectations. The higher-ranking individual is expected to be generous, to protect, sometimes to lead in the face of dangers and difficulties. There is the moral sense of *noblesse oblige*.

Again, Fiske illustrates with the Moose: 'Moose chiefs traditionally "own" all their subjects and, ipso facto, all their subjects' possessions; thus, chiefs can appropriate whatever they like. This principle is so powerful that Moose chiefs avoid entering markets, because vendors would all have to make them gifts from their wares.' But, 'the Moose explicitly acknowledge a major constraint that often operates in [authority ranking] relations: If subordinates have any alternative leaders they can follow, leaders who fail to shelter their followers or who display a lack of largesse may find themselves without anyone to rule over.'[10]

Again, this relationship is not just a matter of resources. In communal sharing, decisions are taken by search for a common path, but in authority ranking, they are handed down through the ranking order. The decisions of the higher-ranking person take the form of decrees to be obeyed. Those lower in rank often watch, copy and defer to those higher; not for nothing do we say that an honoured person is 'held in high regard'. A particularly pure example of this mode for managing decisions is provided by modern military organisations.

Authority ranking, necessarily, is a dominant mode between parents and children, and also conspicuous between children themselves. Fiske suggests that a sense of rank begins around age three and, in groups of older children, ranks rapidly and spontaneously emerge. Across cultures from Bushmen to Europeans, higher-ranking children initiate games, organise the behaviour of others, mediate in conflicts, play with more children, initiate physical contact with protective gestures and are 'more aggressive than average but are not the most aggressive group members'.[11]

Eibl-Eibesfeldt describes fascinating research on children's conflicts, suggesting how the facial expression can be used to predict the winner.[12] In the 'plus face', the head is tilted back, and the gaze directed at the adversary with the inner part of the eyebrows raised. The child showing the plus face will most likely be the winner. In the 'minus face', the chin and eyebrows are lowered and the child looks away. This child will be the loser. In older children, the plus face is also seen as a challenge is met, giving an impression of strength and competence.

This social mode has clear similarities to the rank orders of many social animals, including the pecking orders of hens, jackdaws and other birds, and social ranking in groups of baboons, chimpanzees and other primates. A pecking order has an obvious benefit; when two group members compete, it is far less costly to resolve the dispute not with an actual battle, but with advance knowledge of who would win and who would lose such a battle. This is similar to the advantage of clear signals of relative strength, considered in Chapter 1, but now based on advance knowledge of the individuals involved.

Similarities between human and nonhuman pecking orders, however, go far beyond this. In jackdaws as in humans, the pecking order does not just resolve conflict involving the higher-ranked bird itself. This bird also intervenes in squabbles between two lower-ranking birds, usually in favour of the weaker. The higher-ranking bird is more 'regarded'. 'For example, if a young bird shows fright at some meaningless stimulus, the others, especially the older ones, pay almost no attention to his expressions of fear. But if the same sort of alarm proceeds from one of the old males all the jackdaws within sight and earshot immediately take flight.'[13]

The same thing happens in a baboon troop:

On one occasion when the band was in a treeless area and in danger of encountering a lion, the animals stopped and the young, strong males formed a defensive circle round the weaker animals. But the oldest male went forward alone, performed the dangerous task of finding out exactly where the lion was lying, without being seen by him, and then returned to the horde and led them, by a wide detour around the lion, to the safety of their sleeping trees. All followed him blindly, no one doubting his authority.[14]

In jackdaws and baboons, just as in human communities, rank is closely linked to age, with the responsibility for group decisions accorded to the older, more experienced members.

Like communal sharing, authority ranking comes with its own moral sense. Not only is there the responsibility of the leader to lead and protect, there is the responsibility of the follower to obey. More generally, it is expected that each member of the community will act as their position in the ranking dictates, deferring to those above, and instructing and assisting those below.

In his fascinating 1922 autobiography, of which more later, Henry Ford states confidently that many workers do not want advancement.[15] They are happy with their role, challenges and responsibilities, and do not want more. Fiske considers 'African cultures in which everyone, including slaves, considers [authority ranking] relationships to be natural and inevitable'.[16] In this sense, the authority-ranking mode, like any pecking order, is a strong force for social stability. In *Leviathan* (1651), Thomas Hobbes argued that only an all-powerful authority could prevent human life from collapsing into chaos.

As Eibl-Eibesfeldt points out, for a system like this to work, we need not only an instinct to dominate, but an instinct to submit. These will be found only in social species where such instincts make sense. Any pet owner knows the ease of disciplining the highly social dog, and the near impossibility of disciplining the largely solitary cat. Eibl-Eibesfeldt recounts: 'When I reared a badger I could never forbid it to do anything. If I scolded it when it opened a cupboard and pulled out my linen, then the most it did was to stare at me, and if I gave it a smack on its nose it attacked me.'[17] (Once again, we can only marvel at the ethologist's appetite for keeping animals in the home.)

Interacting in the authority-ranking mode is quite different from interacting in a communal-sharing mode and, once again, the relative emphasis on one or the other varies strongly between contexts and cultures. In the !Kung Bushmen, according to Eibl-Eibesfeldt, there is strong social pressure towards equality. Those who accumulate possessions, or who brag about success, may be punished, criticised or ostracised.[18] Fiske describes societies in which 'no one – however much prestige he or she may have – presumes to tell any other adult what to do'.[19] I sometimes enjoy trying to tell Americans about the very English concept of 'knowing your place'; usually, this leads quickly on to remarks about the merits and inevitability of the revolution.

Authority ranking may involve a linear order, but in human cultures there are many lines. We all know how our rank can change as we move from respected authority figure at work to the despised father of an adolescent at home, or with a switch in the conversation from science to music. To give the last word again to Eibl-Eibesfeldt, 'In the Eipo, a person with a "good garden soul" . . . is esteemed. They even refer to a "sweet potato bigman." Others, who are brave and fight skillfully, are war leaders, while others are unchallenged authorities on construction of houses.'[20] In our complex lives, there are many ranks but, in this social mode, there are strong common themes in how we implement and use them.

Of course, in humans, as in other animals, rank is not fixed. To know your rank is useful, but at the same time, there is a constant element of competition and struggle for position. We shall be coming back to this in Chapter 11.

Equality matching

Fiske calls the third mode *equality matching*. In equality matching, quite unlike either communal sharing or authority ranking, things are carefully matched. It is a matter of fairness and one-for-one exchange, with careful keeping track of who has put in what.

In our own Western culture, we are highly familiar with this mode. We manage dinner party invitations on a strict principle of reciprocity, often alternating between invitations in the two directions. We exchange items of little or no value, such as

Christmas cards, simply respecting the principle that each shows regard for the other. In other cultures, people may exchange brides, precious objects, assistance in agriculture or warfare – and in my own childhood, when we brought in the harvest, neighbouring farmers turned up as a matter of course to help and, next week, when their corn was ripe, we turned up for them. The things that are exchanged can be quite abstract, such as policy concessions between political parties as they struggle to form a coalition. Fiske suggests that equality matching is the guiding principle in democracy, with each person given the precisely equal share of a single vote.

As with the other social modes, equality matching too has its own morality. Here, it is the sense that you should put in as much as you take out, that you should not long remain indebted to others, that things should be fair. Again, notice how different this is from the moralities of other modes – in communal sharing, that we are all one, and in authority ranking, that we know our place, giving and taking according to our position in the scale of things.

Though Fiske suggests that equality matching is seen in every culture, once again its relative weight varies strongly, both between cultures and between individuals. He describes how, 'in New Guinea, Gahuku-Gama soccer teams play as long as necessary, often for days, until they reach a tie'.[21] Many years ago, on a beautiful early summer evening in Cambridge, I met my teenage son after work for a tennis match. On the courts of the peaceful, green Clare College sportsground, we spent an hour belting the ball back and forth between us, striving with every muscle for victory. On the next court, we were dimly aware of two women, obviously great friends, simultaneously enjoying their tennis and a chat, and at the end of their second set, we heard one of them say, 'Shall we play a third . . . or shall we stop now while it's all nice and equal?' They elected for the second option. We thought it was very funny.

Cultural variations in the drive towards equal shares have been studied with a variety of 'economic games'. In the ultimatum game, for example, there are two players who will share a fixed sum of money between them. Player A gets to decide the shares; player B chooses to accept or reject the share they are offered, and if they reject, neither player gets anything. If players were entirely selfish, player A would offer the least possible amount above zero, and B would accept it. Of course, this is not what happens – among Western students, player A most typically offers 50 per cent, and if offers get too low, player B rejects them. The dictator game is similar, except that B has no choice but to accept whatever they are given. Even now, the average amount given away by player A is perhaps 20 to 30 per cent of the total.

In one ambitious study, led by the anthropologist Joseph Henrich and carried out by field workers from Africa to Oceania to South America, games like these were tested in a wide variety of cultures, including foragers, nomadic herders, slash-and-burn horticulturalists and others.[22] In each case, players A and B were anonymous to one another, and these were not games of the imagination; real, large sums of money were on offer. Despite significant variability over cultures, the study found no culture that was strictly selfish. Even in the dictator game, with no possible adverse consequences, and even when the partner is an unknown stranger, taking all the money does not feel right.

According to Fiske, equality matching appears a little later in life than the previous two modes, with children beginning to show this sense of reciprocal fairness around age four. Fiske also suggests that the strict reciprocity of equality matching is uniquely human; no other species keeps exact count in this way. Hints of something similar, however, are certainly seen in other social species, from dogs to crows to primates. In typical experiments, two animals learn to work for a reward provided by the human experimenter.[23] After a while, animal A is still given the reward, but for animal B, something less attractive (though still better than nothing) is substituted. If A were not visibly there getting the better reward, B would keep working, but seeing A's success, B gets angry and refuses. B will not play when the game is unfair.

It seems likely that the urge to reciprocate has played an important role in the evolution of human social life, with our spectacular aptitude for large-scale cooperation. Over the past 50 years (I wrote one of my first undergraduate essays on the question), theorists in biology have wondered how altruism can evolve. Helping one another may be beneficial to both parties, but suppose there is a mutation that makes one party cease to help. Surely, this party now has the evolutionary advantage – he receives the help of others, but gives nothing in return – and genes for altruism must slowly be eliminated from the population. Of course, though, we did not evolve to help others. We evolved to help others who help us. When somebody screws us over, we do not turn the other cheek. The response is a quite clear 'in that case, fuck you, buddy'.

With equality matching and fairness, it is not all roses. Just as we match Christmas cards and votes, we have the principle of an eye for an eye and, around the world, we see the need to match one insult or act of violence with the same in return. Probably no other animal takes revenge in quite this careful, scrupulously 'fair' way either. This is just us, controlled by another of our own, powerful social modes.

Market pricing

Fiske's fourth mode dominates a great proportion of our modern life, but I don't have much to say about it here. This mode he calls *market pricing*. In contrast to equality matching, where specific things are exchanged one-for-one, in market pricing there is a universal scale of value. This means that we can say how many apples have the value of a loaf of bread or a political favour, opening the door to a much broader range of social exchanges.

Because we have money in modern civilisation, this form of social exchange is everywhere. Money makes values and market pricing very explicit, and provides a universal mechanism for this kind of social management. Fiske argues that this mode too has its innate origins, and it is plausible that, even before the invention of currency, humans were perfectly able to compare the exchange value of different things.

I will be coming back to money in a later chapter, but for now, it seems hard to say that this fourth mode has its own, unique morality, or sense of what is right and wrong. Rather, our moral views concerning money seem to be built on values we have already seen – that we should not cheat others, that we should only take what we deserve, that we should use our resources to support those less fortunate, and so on.

Combining the modes

If these social modes are innate releasing mechanisms – inherent blueprints of how we should behave, filled in with the countless specifics of individual cultures – they have all the IRM's usual conflicts and inconsistencies. In communal sharing, we put indiscriminately into the common whole; in equality matching, we keep things fair; in authority ranking, we accept that some are given more than others. These moralities cannot all apply at once; instead, the one that predominates varies across time, context and culture.

In some cultures, the relationship between a husband and wife is one of authority ranking; the husband expects to dictate, the wife expects to obey. In other cultures, it is one of equality matching; for many years, I have held that the key for a successful marriage is for both parties to be trying to put in at least as much as they take out (though, strictly speaking, this is more like market pricing, since a cup of coffee brought to bed in the morning can approximately match picking up a child at the end of the day). Even within one relationship, the two can be jumbled up. Coming back from school one day, my younger son, perhaps aged five, told us he had been arguing about families with his friends in the playground. In all delightful innocence, he told us, 'They all said it was dads who were in charge, but it's mums, isn't it?' (No comment.) We explained that we didn't think of it that way, that everybody has a say. It may be doubted whether this parental ploy successfully replaced authority ranking with communal sharing in his developing social awareness.

Also like other IRMs, these social modes have their own intrinsic force, with a spontaneous need to be discharged and a sense of felicity when they are satisfied. The need for equality matching means that a relationship struggles if either partner takes out more than they put in; there is a sense of resentment if we receive too little, but a definite discomfort also if we receive too much. The authority-ranking mode brings the joy of receiving the captain's armband or a word of commendation from a superior, but also the comfort of 'knowing one's place', wherever that may be. The modes tell us how to behave and, as they are discharged, they bring a sense that our social lives are right.

More morals

Fiske's modes underpin different facets of human morality and, of course, there are likely to be more. Let me illustrate with another proposed scheme, the 'moral foundations theory' of the psychologist Jonathan Haidt.[24] We talked briefly in Chapter 3 about Haidt's moral vignettes, and how they are used to illuminate intuitions about right and wrong, again with strong common elements across very diverse cultures. Haidt studied under Fiske, and his view of fundamental moral modes shares some strong elements with Fiske's, but with significant additions. He calls each fundamental aspect of morality a 'foundation', each named by its opposite dictates of what should be done (moral) and what should not (immoral).

Overlapping with Fiske's communal sharing, Haidt proposes a fundamental 'care/harm' foundation. This dictates that we should look after others (moral) and avoid harming them (immoral). Like Fiske, Haidt suggests that a primary origin for this mode comes from parental care, expanded through evolution to encompass care in general. It is easy to produce a vignette that violates care/harm. If you write a story about a man who sees an injured dog by the side of the road and goes over and kicks it, pretty much everybody will agree that this is wrong – whether or not they like dogs themselves.

Fiske proposed that, in communal sharing, there is a strong sense that the group is threatened by contamination. 'Our people' have an essence, and it is important to protect its purity. On a similar note, Haidt proposes a 'sanctity/degradation' foundation, though he traces the origin of this to disgust and the instinctive rejection of pathogens. People will easily judge disgusting things not just to be disgusting, but morally wrong.

Haidt uses a vignette involving a man who buys a chicken, has sex with it, then cooks and eats it. No harm is done, there are no victims, but there remains a sense that this is not just unpleasant, it is a moral violation. Though this foundation may begin with food, the thought is that, through evolution, it has been coopted into a broader sense of what our community 'does' and 'doesn't' do. It takes a concrete form in precious relics to be worshipped and kept pure, in rules of behaviour such as chastity, in modern semi-religious beliefs that decent people eat organic foods and keep their bodies clean of toxins. If somebody urinates against the side of the church, this is clearly more 'wrong' than urinating against the side of the supermarket. I would imagine that, in many parts of the world, urinating against the side of a religious building would be a very dangerous thing to do.

Haidt has an 'authority/subversion' foundation, very much taken from Fiske – the morality of correctly fulfilling your role within the hierarchy. He also has a 'fairness/cheating' foundation, very much resembling equality matching though without Fiske's specific emphasis on exact one-for-one exchange. Fairness/cheating, in Haidt's scheme, brings up another aspect of human morality, the question of ownership. The person who has something can be taken as its owner, and it is wrong to steal. Eibl-Eibesfeldt thought that the sense of ownership is itself innate, and underlies the bonding power of giving and receiving gifts. As a hint of something similar in other great apes, Eibl-Eibesfeldt describes a film of the primatologist Diana Fossey observing gorillas in the wild.[25] Diana Fossey is lying on her stomach taking notes. A gorilla approaches, takes the pencil, moves away and examines it. Then . . . he returns, gives it back (!) and takes the notebook. Now the performance is repeated – the notebook is examined, then again returned. As Eibl-Eibesfeldt notes, 'baboons and macaques would never behave this way'.

A fifth foundation in Haidt's scheme is 'loyalty/betrayal'. Fiske, perhaps, ignores this because his scheme is not very concerned with conflicts between one group and another but, as soon as we think about it, we know that loyalty to our group is a very strong moral imperative indeed. In Chapter 3, we already saw Haidt's vignette about the woman who cuts up an old flag for rags, and the contorted rationalisations that people use to explain why this harmless act is wrong. The truth, however, is that it is wrong not because it harms anybody at all, but because it is disloyal to the tribe. There are few more scornful nouns in English than 'traitor'.

A sixth of Haidt's foundations, 'liberty/oppression', is interesting because it conflicts very directly with authority/subversion. In the authority/subversion mode, it is right to respect authority, order and the law of the land. This is the morality of Thomas Hobbes and *Leviathan*. But it is also right to object when authority goes too far, and this is the morality of liberty/oppression. Once again, we are in the world of conflicting IRMs, balancing what can be opposite needs in a complex social reality.

Like Fiske, Haidt thinks that each moral foundation is an instinctive fixed point, built into the human mind by the requirements of our cooperative social existence. But they also vary in their relative weights and in where they should be applied. One good example is the increasing sense of animal rights in Western culture – something quite incomprehensible at other times or in other parts of the world.

As a more extreme example, I remember that, when my teenage responsibilities included chopping firewood, I would sometimes be engaged in a mighty struggle with a particularly knotty log, and suddenly there was a slight sense that it was unfair to destroy such a worthy competitor. Roald Dahl has a rather creepy version of the same thought in 'The Sound Machine' (1949), his story of a man who invents a machine to hear the screams of plants as they are cut down.

An especially interesting part of Haidt's thinking concerns differences between political ideologies. Thinking of the political turmoil of the current United States, it is not hard to see the differing weights given to care/harm, authority/subversion, liberty/oppression and the rest. If we think of ourselves as one, and that poverty is misfortune, then we are in the mode of care/harm and, of course, we support welfare and a universal basic income for all. If we think of the right of possession and fairness/cheating, then our money is taken unfairly in tax, and what we have earned by our efforts is given to undeserving freeloaders. If we think that our nation is threatened, it is right to give everything, up to and including our lives, for the loyalty/betrayal foundation, but if we think that government is against us, it is liberty/oppression.

Haidt argues that, if we understood this better, perhaps some of the animosities of rival parties could be defused. Perhaps, those under the influence of a different foundation would seem at least comprehensible. Going back to the Delphic oracle, it is 'Know thyself' again.

Blame and shame

Moral principles keep things together. They provide a system of rules allowing the relatively smooth progress of life in a human society. They have two sides. So far we have been focusing on the internal side – the force of these principles in giving each person a sense of correct behaviour. But there is also an external side, as the same principles are used to control the behaviour of others – through disapproval, shame, mockery and sometimes violence if principles are disobeyed. Haidt and others argue that both sides were important to allow morality to evolve. Not only do we need a sense of right behaviour in ourselves, we need a community that will gossip, criticise and blame when the rules are broken. Darwin says the same in *The Descent of Man*. He argues that, as instincts towards self-gratification are balanced against the needs of others, a major force in the decision is the need to 'avoid the disapprobation, whether reasonable or not, of . . . fellow men'.[26]

Analysis of everyday conversations shows prodigious amounts of time devoted to gossip, often disapproving.[27] It is hard to imagine a culture without practices to ensure that moral imperatives are obeyed. Our moral IRMs thus have two sides. Releasers such as a friend in need of support, or an undeserved demand from authority, trigger the matching behaviour of care or rebellion. Complementarily, the perception of a group member who is not following the rules releases shaming and blaming.

Blueprint and detail

Fiske's four social modes, or Haidt's six moral foundations, capture broad constraints on the instinctive management of human relations. Spock fills them in with particular cultural details, so that a particular conch shell becomes treated as a sacred object of exchange, or neighbours lend work and machinery for a modern corn harvest. We may think that the specialised, social blueprints from Pan contain open slots – 'insert any valued item here' – that can be used to turn the blueprint into concrete moral code.

Completed with cultural details, Pan's blueprints complement the many, more specific IRMs that also govern our interactions – the specific urge to pick up and cuddle an infant with large eyes and bulging forehead, the urge to bow before a superior, and much more. When we consider the richness and complexity of human relationships, it might seem astonishing that heavily innate constraints can make things work at all. But we know how elaborate are the physical structures produced by evolution, from the spectacular macroscopic design of an eye to the countless cellular and biochemical processes that allow it to develop from an embryo and then to function in the adult. Against this background, it seems less surprising that evolution can also produce elaborate, finely crafted structures of behaviour.

As Spock adds details to Pan's blueprints, there is all his usual focus, favouring one moral mode over another. On the one hand, 'children should be seen and not heard' is a cultural completion of open slots in the authority-ranking mode. On the other hand, this strict, abstract, highly limited rule is a clear violation of everything essential in communal sharing. Despite the common themes of Pan, across cultures there are very wide variations in the sense of correct interaction between adults and children.

Now, we need to complete the picture we have started. We need to consider that these elaborate, finely crafted structures of bonding and social interaction evolved in a context of small bands of hunter-gatherers. For groups of territorial animals, it is not all bonding and making the group work. As Lorenz tells us, the other side of the story is joint territory defence, and here too, we do it in our own, equally complex and extremely dangerous human way.

CHAPTER 8

Them

Defeating the enemy

In part, as we have seen, animals form groups to strengthen the team. Members of the same species are in competition for territory and resources, and in this competition, the team carries more weight than the individual.

Lorenz gives a rather appealing example for the greylag. Though the triumph ceremony usually bonds a mating pair and their family, sometimes it is two males who end up bonded in this way. Again, the bond is not primarily sexual, though there may be initial failed attempts to mate. It conspicuously is, however, about mutual support, and because male geese are physically stronger than females, a bonded pair of males rapidly rises up the pecking order. In the fourth season of *Peaky Blinders* (2017), Tom Hardy's criminal boss gives his summary of life: 'Big fucks small.' He is a criminal boss, and as we have seen in the last two chapters, his view is one-sided. From greylag geese to social insects, however, animals come together to give 'us' a better chance against 'them'. Innate releasing mechanisms for bonding and cooperation are intimately bound up with IRMs for attack and defence.

This pattern is strongly seen in apes and other primates. If two groups of baboons forage into the same region of an acacia forest, there is a stand-off. Usually, one group abandons the area but, if not, things can escalate into battle. In Uganda, bands of male chimpanzees march along the boundary of their group territory, on the lookout for invaders. They appear vigilant, sometimes climbing trees to peer out, sometimes pausing at an overlook. If chimps from another territory are encountered, war breaks out. Plants are shaken, there are loud cries, rocks are thrown. Sometimes, the rivals are attacked and killed. Chimpanzees may also raid a neighbouring territory, killing the males and stealing the females.[1] Eibl-Eibesfeldt reports that, when the Yanomami are asked why they go to war, they explain that they want to steal women.[2]

As always, Eibl-Eibesfeldt makes much of detailed similarities between humans and other primates. Here is his account of territory defence in vervet monkeys:

> When a vervet monkey troop forages on the ground, several males will 'stand guard' with their backs to the troop. They spread their legs slightly and display their colorful sexual organs. In these monkeys the scrotum is blue and the penis is bright red; here signal effect appears to have been selected for. The guarding is directed towards conspecifics of other vervet monkey groups, and if they approach too closely, the guards develop erections . . . This behavior can be interpreted as a ritualized threat mounting. Mounting is used as a sign of dominance in many mammals, and as such, is detached from its original function of copulation.[3]

He then proceeds with photographs of our own use of phallic displays to deter threat – figures from Bali, crouching, grimacing, with erect penis, used to ward off evil spirits; amulets from Japan, with a threatening face on the front and a penis hidden in a compartment at the back; an Eipo man, with a two-foot fake erection thrusting out from under his loincloth, his silhouette exactly resembling the silhouette of a rock guitarist erecting his instrument.

In the modern world, we are more hesitant about this, but as Eibl-Eibesfeldt points out, a similar link of sexuality and aggression can still be traced in otherwise incomprehensible verbal expressions: in English, 'Fuck you,' and apparently among Arabs, 'Phallus in your eye.'

The tension between us and them is all around us. It is only a slight simplification to say that it is a tension between good and evil. When we treat people as 'us', there is a strong, sometimes overwhelming sense of good. We feel evil when somebody else treats us as 'them', and in some way attacks us. We also feel evil when somebody else is attacked – as long as this is somebody with whom we can sympathise. When we treat a perceived enemy as 'them' . . . well, now it is not so simple. As Darwin put it in *The Descent of Man*:

> The virtues which must be practised, at least generally, by rude men, so that they may associate in a body, are . . . recognized as the most important. But they are practised almost exclusively in relation to the men of the same tribe; and their opposites are not regarded as crimes in relation to the men of other tribes. No tribe could hold together if murder, robbery, treachery, &c., were common; consequently such crimes within the limits of the same tribe 'are branded with everlasting infamy'; but excite no such sentiment beyond these limits.[4]

Or here is Lorenz, likening a palaeolithic tribe to a modern fighting unit:

> We know to what heights of heroism and utter self-abnegation average, unromantic modern men have risen under these circumstances. Incidentally, it is quite typical of man that his most noble and admirable qualities are brought to the fore in situations involving the killing of other men, just as noble as he is![5]

In *The Blank Slate*, Steven Pinker pillories the rather idealistic belief that war is an invention of civilisation, a corruption of 'the noble savage'. Using data from the archaeologist Lawrence Keeley, he gives a rather telling bar graph showing the percentages of male deaths caused by warfare.[6] In a range of indigenous groups, from South America and New Guinea, the percentages range from 10 to around 60. In the US and Europe, throughout the twentieth century and including two world wars, the number is less that 1 per cent.

In *Human Ethology*, Eibl-Eibesfeldt details tribal conflicts around the world, from organised battles to surprise attacks on neighbouring villages.[7] Sometimes just the men are killed, sometimes women and children too. Sometimes the dead are eaten, just as chimpanzees too will eat the bodies of their victims. Sometimes things begin with a minor affront and then escalate into widespread warfare. Sometimes violence concerns territory or other resources; the battle is for water holes or grazing lands, and the losers are driven out of their territory. In the Enga of Papua New Guinea, encroachment on land was cited as the cause for war 'twice as frequently . . . as pig stealing or murder'.[8] One study of the Yanomami found that 44 per cent of men aged 25 or above had participated in killing another person. Battle is depicted in cave and rock paintings from the Kalahari to Europe, with all the drama of men running, clashing, pointing their weapons and falling, pierced with arrows.

War is the extreme but, even without war, the tension between 'us' and 'them' is a deep, spontaneous principle of the human mind. In social psychology experiments, groups can be formed by the most absurd means, perhaps giving one group red shirts and another blue, and, immediately, the participants want their own group to win the advantage.[9]

A few years ago, my wife and I found ourselves in the 'away fans' stand, watching a football match between our team, Liverpool, and the home team, Aston Villa. In almost the last minute of the match, Liverpool scored an unlikely winning goal and the stand around us exploded, with bellowing men pouring in a straining mass to the front, waving triumphant fists at the Villa fans, fingers raised in the modern Western form of the phallic threat, and as the stewards asked my wife if she was OK in the melee (needless to say, she was more than OK), I felt certain that, had there been broadswords and no barriers, the swords would have been swinging. In the lengthy, painful Brexit wars, I do not believe that finally it was about economics, or protection of borders, or even national sovereignty; it was about the opposing teams of Brexiteers and remoaners, and when Jeremy Corbyn tried to fight an election without saying which team he was on, he and his Labour Party were crushed.

In the 'us' and 'them' wars, Spock is remarkably powerless against the strength of Pan. Within our own group, the sense of 'us' is so strong and natural that we cannot really grasp its irrationality; from any outside perspective, it seems ridiculous. When Tom Hanks famously outed his childhood drama teacher at the Oscars, chokingly describing him as one of the 'finest gay Americans', from the other side of the Atlantic I could only think,

Why on earth would you have a concept of gay Americans? Are they
somehow finer than the gay Ugandans?

In the first Gulf War, according to the British Legion website,
392 coalition troops were killed, including 47 British. These 392
can be set against an estimated 20,000 to 35,000 Iraqis. Spock
knows, as Lorenz knew, that Iraqi men are just the same as
British, French, Canadian and American men, with all the same
aspirations, fears, nobility, children. But he can never get the
message across. On the Highway of Death, when fleeing Iraqi
troops were strafed and bombed as they tried to make it back
to Basra, the American journalist Kenneth Jarecke photographed
the scorched corpse of an Iraqi man. The eyeless face and straining body suggest a last, agonised struggle to escape from the
burning truck. Jarecke later explained that he thought, 'If I don't
photograph this, people like my mom will think war is what they
see on TV.' As it turned out, though, the American media were
more than happy to protect moms from his image, which is now
famous but at the time was little seen. Wartime TV shows what
Pan finds it natural to be shown: a handful of coffins draped in
Union Jacks or Stars and Stripes, with a grave voice speaking of
'lost American' or 'lost British' lives.

I have always been rather fond of this remark made by Herodotus, the Greek 'father of history' writing in the fifth century
BC. Herodotus had travelled widely through the known world,
from Egypt to Babylon to southern Italy, and in *The Histories*, he
synthesised what he had learned of places, peoples and events.
Cosmopolitan, thoughtful and humane, Herodotus knew quite
well the foibles of Pan:

> If anyone, no matter who, were given the opportunity of
> choosing from amongst all the nations in the world the set
> of beliefs which he thought best, he would inevitably – after

careful considerations of their relative merits – choose that of his own country. Everyone without exception believes his own native customs, and the religion he was brought up in, to be the best.[10]

The other side of bonding

In Chapter 6, we considered the forces that can produce a welling sense of 'us' – laughter, music, dance, sharing. But every force for the togetherness of 'us' is also a force for the exclusion of 'them'. By definition, if there is a group, some people are inside it, and others are not. We have laughing together – and we have laughing *at*. We have shared music – and in our culture, each age group uses its own musical genre to separate itself from the previous generation. I reached my teens in the 1960s and, to me, the golden era was the Beatles, the Rolling Stones, Bob Dylan and Janis Joplin at Monterey. When I show a clip of Janis taking rock and roll to a new place in 1967, younger friends think it is annoying screeching.[11]

By analogy with biological evolution, the psychiatrist Erik Erikson coined the term 'pseudospeciation' to describe the formation of cultural groups.[12] The group adheres rigorously to shared beliefs, customs, habits and values; those with different beliefs and habits are rigorously excluded and avoided. In groups of children, it is especially clear that pseudospeciation is not optional. If a child wishes to belong to the group, then it is obligatory to adopt its beliefs, its music, its values and mannerisms. The child who is different will be persecuted and mocked in a way Lorenz likened to the mobbing of predators in birds and other animals.[13]

Shared habits and beliefs extend from the most major aspects of a person's life, perhaps their sacred religious beliefs, to the most minor details of 'good manners', smoothing relations when they are shared and causing offence when they are not. Here is a typically Lorenzian example:

A good example is furnished by the attitude of polite listening which consists in stretching the neck forward and simultane-ously tilting the head sideways, thus emphatically 'lending an ear' to the person who is speaking. The motor pattern conveys readiness to listen attentively and even to obey. In the polite manners of some Asiatic cultures it has obviously under-gone strong mimic exaggeration; in Austrians, particularly in well-bred ladies, it is one of the commonest gestures of polite-ness; in other central European countries it appears to be less emphasized. In some parts of northern Germany it is reduced to a minimum, if not absent. In these sub-cultures it is con-sidered correct and polite for the listener to hold the head high and look the speaker straight in the face, exactly as a soldier is supposed to do when listening to orders. When I came from Vienna to Königsberg, two cities in which the difference . . . was particularly great, it took me some little time to get used to the polite listening gesture of East Prussian ladies . . . I could not help feeling I had said something shocking when she sat rigidly upright looking me in the face.[14]

Shared habits give a sense of understanding and, indeed, we use understanding itself to create the sense of a common 'us'. When a miserable friend is crying about their lost partner or sick child, very often we extend sympathy with an account of our own similar experience and shared sadness. Eibl-Eibesfeldt points out that, across human languages, intimacy is created with state-ments of agreement and common understanding, sometimes even as trivial as, 'Yes, the weather certainly is beautiful today.'[15]

Many years ago, my older brother, always rather a positive individual, told me how he had often had trouble getting on with an electrician who sometimes worked in his business, and who always appeared to be in a bad mood. One morning, as they stood dripping in the pouring rain, and for some reason, my brother was having a dismal day, the electrician asked my brother how he was and he replied, 'Fucking awful morning.' Ever after, they were friends.

Often, despite the complaints of philosophers through the ages, we confuse understanding and acceptance. When the psychologist traces the causes of a criminal's act, this seems somehow to justify and forgive, though, rationally, understanding and acceptance are two quite separate things.

Complementarily, we degrade 'them', keeping ourselves blind to their essential similarity to ourselves. Often 'they' are treated as animals, not human at all. I watched with deep shock when Donald Trump described illegal immigrants as 'animals', but this method is everywhere, from Nazis speaking of racial hygiene, to the untouchable castes of India, to the Eipo who, according to Eibl-Eibesfeldt, describe their enemies as 'dung flies, lizards or worms'.[16] Justifying his actions over the Highway of Death, General Norman Schwarzkopf explained, 'This was a bunch of rapists, murderers and thugs' – though, of course, the bomber never knows exactly who his bomb will hit. An intriguing study by psychologists Gordon Hodson and Kimberly Costello found that people with the most negative attitudes to immigrants or other out-groups scored high on measures of interpersonal disgust – for example, reluctance to sit in a chair still warm from another person's rear.[17]

The sense that somebody is with us or against us, one of us or one of them, explains something quite irrational in the human mind. We judge people as wholes, but of course they are not – each person is an individual, with many beliefs, many actions good or evil. It is common to wonder at the Mafia boss who has his enemies strangled but is kind to children and animals, or the great artist who exploited women, but in the complex, incoherent world of Pan and Spock, there is nothing to wonder at. As we have seen, Lorenz the thinker could also be Lorenz the Nazi. For insight into ourselves, it is the ideas of Lorenz that we need, and to Spock it doesn't matter where a good idea comes from. For Pan, though, it matters a lot.

Opposing division

In many species, we have seen the conflict between affiliative and aggressive IRMs. In ourselves, the modes of 'us' and 'them' have the same tension and, in the blink of an eye, we can switch between them, very much like any other animal switching between competing IRMs. Emerging from a football match in the 1970s, swept along by chanting skinheads in red and white scarves, I found myself behind another man struggling to protect his small son in the unbelievable crush of bodies. The skinheads around me, who throughout the match had been raining a storm of spittle on to the police around the perimeter of the pitch, were saying, 'Are you OK, mate? Can we give you a hand?' Then one of them spotted another man, also swept along with us, in the blue and yellow of Oxford. I heard one of them say, 'What's he – Oxford? Kick him!' and they surged away to do it.

Eibl-Eibesfeldt describes his rage when an overtaking driver cut in on him, rapidly changing to a cheerful wave when the offender was recognised as a friend; and an American soldier whose life was saved when he came face to face with two soldiers of the Vietcong, his gun misfired, and he smiled.[18] These ideas did not begin with the ethologists. Eibl-Eibesfeldt quotes this passage from Freud:

> If willingness for war is a consequence of the destructive drive, then the answer is to call upon love (Eros), the opponent of this drive. Everything that creates emotional bonds between men must be pitted against war . . . Everything that establishes significant points in common between people arouses . . . fellow feelings . . . On these to a great extent rests the structure of human society.[19]

The idea that we might use the forces of unity to oppose the forces of opposition and prejudice was most importantly put forward by the psychologist Gordon Allport in the 1954 book *The Nature of Prejudice*. Studies during and after the Second World War had begun to suggest that simple contact between groups promotes understanding and acceptance . . . very much in line with Freud's comments.[20]

The more voyages that white seamen took with Black seamen working alongside, the more positive to Black people they became. White police officers who had worked with Black colleagues found it easier to accept the authority of Black superiors. In ambitious semi-experimental studies in New Jersey and New York, families were assigned to segregated or desegregated housing blocks. White women in the desegregated blocks had more positive attitudes to both their Black neighbours and to desegregation in general.

Based on early projects like these, Allport proposed that organised contact can be used to break down the barriers of us and them. To make this work, he thought, contact needed the right conditions. As important factors, he included equal status for the two groups involved, common goals and cooperation, and a supporting institutional context.

Allport's ideas led to an enormous research literature, with many findings pro and con. Fifty years later, the psychologists Thomas Pettigrew and Linda Trapp carried out a massive statistical synthesis of this work.[21] Their final data came from 515 individual studies, with a total of 250,089 participants. Many studies had examined prejudice based on race, but there was much more than this, with out-groups defined by sexual orientation, mental or physical disability, age or mental illness. Samples had been examined across the world – US, Europe, Africa, Asia, Latin America, Australia. With the most careful statistical methods for estimating the average effect across all these studies, Pettigrew and Trapp showed that contact does work. This is not to say that it is transformative – the effects are modest rather than huge – but it does work. Just as Freud thought, when we know other people, it can be easier to understand how very much they are like ourselves.

Contact does not just work for race – the average effect is very much the same for all kinds of in/out-group prejudice. Allport's ideas about necessary conditions do not stack up so well against the facts – in Pettigrew and Trapp's analyses, there was little evidence for the importance of Allport's individual factors, though some evidence that contact works best in the setting of an organised, structured programme. Perhaps most important, the effect generalises. It is not just that contact promotes tolerance between individuals – increased tolerance spreads to the entire out-group. Most striking of all, it also spreads to new out-groups that were not even involved in the study itself. Coming to understand members of a previous out-group has not just made the participants like them better. It has made these participants better people, more open to the world of others.

Lorenz also believed that aggression between groups is

defused by contact, friendship and shared values – by all those forces, indeed, that forge a group of 'us'. He put it like this:

> No one is able to hate, wholeheartedly, a nation amongst whose numbers he has several friends. Being friends with a few 'samples' of another people is enough to awaken a healthy mistrust of all those generalizations which brand 'the' Russians, English, Germans, etc., etc., with typical and usually hateful national characteristics.[22]

He goes on to speculate, rather winningly, that it would be impossible for him to wish for the destruction of an enemy country 'if I realized that there were people living in it who, like myself, were enthusiastic workers in the field of inductive natural science, or revered Charles Darwin'.[23] This, from Lorenz the erstwhile Nazi. That is the patchwork and the chaos of conflicting IRMs writ large.

The shifting balance between modes of 'us' and 'them' is shown in one of the best-known studies in social psychology, now around 70 years old.[24] These were the 'Boys' Camp' experiments, carried out by Muzafer and Carolyn Sherif in the early 1950s. Each year, groups of ordinary American boys attended an experimental summer camp. They were divided at random into two groups, housed in different bunkhouses. After a beginning period of harmony, the experiment began. It had two phases.

Phase I – 'them'. The experimenters set up a series of games and competitions between the rival bunkhouses. There were points, winners, prize knives and medals . . . everything was done to make each group want to come out on top. Very rapidly, relations deteriorated to the point of juvenile war. There were raids on the enemy bunkhouse, with beds destroyed and precious items stolen. There were flag burnings and pitched battles, with furious boys calling the opponents to come out and fight. Members of the other bunkhouse were considered 'smart alecks, stinkers'. Boys refused to touch things that belonged to the other group. At mealtimes, the groups colonised different tables; there were insults, catcalls, objects thrown at the opposition. Only interventions from the staff prevented outright violence.

Phase 11 – 'us'. Perhaps unwilling to send the boys home corrupted, the experimenters introduced Phase 11. Now, activities called for cooperation and shared goals. There was a water shortage – artificially created by the experimenters, but real enough to the boys, who soon had little left in their drinking canteens. After a long day of increasing concern, both groups of boys were set together to find and clear the artificial blockage in the water system – which they successfully did together, sharing activities and tools. On another day, a truck that was needed to bring food apparently would not move. The food was needed . . . the boys realised that working together they might attach their tug-of-war rope and haul the truck forwards until it started, even chanting together 'heave . . . heave . . . '. There was more along the same lines, and very soon, hostility was gone. It was back to normal boys, hunting lizards together.

In our war-torn world, it is easy to see why these experiments have left such a lasting impression. These were only psychologists' games and innocent boys, but the results show ourselves – more than willing to fight when it is a matter of competition, more than willing to help when we are in it together. The opposing modes of us and them, vying for control of our minds.

No 'us'

The idea that heavily innate 'us' and 'them' modes compete for control raises an obvious possibility. Very likely, the strength of these two competitors will vary from person to person. Surely we know that some people are more friendly, some more aggressive. In some people, perhaps, the competition is so unbalanced that a 'them' mode is always in control.

Many years ago, as a postdoc in Oregon, I went with a group of friends to see Andy Warhol's *Bad* (1977). *Bad* is a quite brilliantly executed study of simple, unemotional, transactional evil. A group of women will do anything they are paid to do. You want to be rid of your autistic son? – pay up, and your problem will disappear. Your neighbour has insulted you? – his dog can pay the price. It is all done without concern, without hatred, without anything except company policy. I basically never walk out of films but, by the end, we all wished that we had.

Outside the movies, the psychopath is real – cold, manipulative, callous, with no care for others or sense of their suffering. Though the clinical names for this condition tend to change, the reality is clear enough. In the psychopath, relationships are shallow and one-directional, with little desire or ability to form relationships based on mutual love and caring. There is a sense that other people are objects, to be used as they are needed. In Fiske's terms, the communal-sharing mode seems to be missing. The psychologists Essi Viding and Eamon McCrory suggest that, for the psychopath, there is 'an extremely restricted in-group: themselves'.[25]

As we might expect for an imbalance between IRMs, psychopathy is heavily heritable. There is some evidence for a link to genes that affect the release of oxytocin – a hormone involved in social bonding across many species. In Chapter 6, we saw the role of shared laughter in bonding and, in one study, Viding and her colleagues showed that boys at risk for psychopathy show a reduced tendency to laugh along with others, and altered brain responses to the sound of laughter.[26] Other studies have measured eye contact between these at-risk children and their mothers, with fascinating results: while the mother seeks contact by looking at the child, the child, rather heartbreakingly, does not look back.[27] It is as if, in these children, the deep joy of simply being with a loved one is missing. Intriguingly, callous and unemotional behaviour is also sometimes associated with extreme childhood adversity but, here, Viding argues, there is a different psychological profile, with extreme response to anything that could be seen as a threat.

This is not a difficulty with understanding the minds of other people. Psychologists have devised many tests to ask how an adult or child 'mentalises', or takes another's perspective. Children at risk of psychopathy can understand another's perspective perfectly well. Correspondingly, the adult psychopath can often understand others all too well, as those others are treated as objects to be manipulated. It is not the understanding that is missing. It is the sense of 'us'. It is upsetting precisely because it violates our sense of human good.

Does it have to be this way? Another intriguing question is posed by the idea of competing mental modes out of balance – is the 'us' mode missing, or simply so weak that it never comes into control? Remember those experiments with mice, and the area in the ventromedial hypothalamus that releases aggression. Mice, like people, vary in how aggressive they are, and some gentle male mice will never attack when another is introduced to their cage. But this does not mean that the behaviour pattern is not built into their brain – stimulate the ventromedial hypothalamus, and the previous pacifist mouse springs into a fully formed, normal attack.[28] The attack mode was there all along, but hidden until an artificial stimulus brings it out.

Of course, we respond to the psychopath with punishment, with discouragement of the 'them' mode, but in a last strand of

Viding's research, the problem is seen from the opposite perspective. If the 'us' mode is simply too weak, can it be strengthened, perhaps by intervening early in life as the personality is consolidating? As we saw in Chapter 6, we know of many bonding forces, from shared music and dance to shared goals and values. Is there an 'us' mode hidden in there, perpetually kept down by the strength of 'them'? Is it really true that there is some good in everybody?

Loyalty

The greylag has the triumph ceremony, uniting friends and family by battle against an invader and, very obviously, *Homo sapiens* is no different. Nothing perhaps brings a society together so strongly as the threat of a common enemy.

I was born just a few years after the end of the Second World War, and having lived through the time of Dunkirk and the Blitz, my parents would speak of a quite extraordinary sense of national solidarity, of self put aside for the needs of the whole. Over a hundred years after the end of the First World War, Britain still takes a breath one day a year to wear red poppies and remember the dead.

Lorenz points out that two enemies can easily be brought together when an even more threatening enemy appears from outside. It is easy to see this in ourselves, when a nation goes to war and suddenly its own warring political factions are united. Lorenz describes the same phenomenon in his greylag colonies, when an autumn migration brings large numbers of birds back from a distant site. 'Faced with these utter strangers, the otherwise mutually hostile families of geese on our lake unite in one collective phalanx of converging necks, and attempt to drive away the intruders.'[29]

Humans, of course, hunt and fight in packs, and when we do this, we are very much like other pack animals. It is always striking that even a tiny dog will fly at a much larger adversary – something that a cat will never do unless it is thoroughly trapped – but while the cat is solitary, the dog is a pack animal, and even when there is no pack, it attacks as though there were. Pack animals may have specific rituals used to draw the pack together for a fight. Lorenz describes how, when a pack of wolves prepares to attack some vulnerable prey, 'all the members of the pack suddenly gather together and indulge in a ceremony of general nuzzling and tail-wagging'.[30] Teenagers take many more risks when they are together in a group,[31] and having once been a teenage boy myself, it is easy for me to remember the delighted sense of shared optimism, challenge and immortality. It is like the Enga of New Guinea, who, according to Eibl-Eibesfeldt, initially go to war as if it were a game, then later, when somebody is killed, wonder how they could have got into this.[32]

The pack is in it together, and as we saw earlier in the moral imperatives proposed by Jonathan Haidt, just as the pack offers the joy of shared challenge and victory, it demands loyalty. For young boys becoming men, many cultures have demanding, painful, sometimes dangerous initiation ceremonies, often including severe instruction from their elders. Eibl-Eibesfeldt thought that these rites serve to imbue a group spirit, a sense of commitment and loyalty, a willingness to take risks and, if necessary, to die for the common need.[33] Haidt points out the universal scorn expressed for traitors – from Allah, who vows that those who have abandoned the faith will be roasted, to Dante, who condemned traitors to the innermost circle of Hell.[34]

Lorenz thought that, in human beings, the act of coming together against a common enemy produces a very specific, very dangerous psychological state. He called it 'militant enthusiasm', and in the twentieth-century world (now twenty-first), he thought militant enthusiasm was perhaps the greatest danger to the survival of our species. He put it like this:

> One soars elated above all the ties of everyday life, one is ready to abandon all for the call of what, in the moment of this specific emotion, seems to be a sacred duty. All obstacles in its path become unimportant, the instinctive inhibitions against hurting or killing one's fellows lose, unfortunately, much of their power. Rational considerations, criticism, and all reasonable arguments against the behaviour dictated by militant enthusiasm are silenced by an amazing reversal of all values, making them appear not only untenable but base and dishonourable. Men may enjoy the feeling of absolute righteousness even while they commit atrocities. Conceptual thought and moral responsibility are at their lowest ebb. As a Ukrainian proverb says: 'When the banner is unfurled, all reason is in the trumpet.'[35]

We need this too

Of course, we need a 'them' mode because it will not always work to turn the other cheek. When an inspired terrorist pulls out a knife on a crowded London bridge, or an alienated teenager breaks into a schoolroom with a gun, we need the hero who may die to protect the group, and there is good biological reason why the hero's courage releases the most intense sense of admiration and honour. Life will not always leave us alone to be gentle saints, and much though we need the welling sense of 'us', with the enemy at the gate, we also need the welling certainty that we must fight back.

As Louis de Bernières puts it in *The Autumn of the Ace* (2020), 'When a soldier comes through your door with the intention of cutting your balls off and raping your sisters, believe me, you suddenly stop being a pacifist.'[36] Or Darwin in the *Descent of Man*: 'although, in civilized countries, a good, yet timid, man may be far more useful to the community than a brave one, we cannot help instinctively honouring the latter above a coward, however benevolent.'[37]

Furthermore, we do not need balls ripped off and sisters raped to see how these same behaviour patterns surface in every corner of our lives, allowing us to face fears and challenges of all kinds. When I was a small child, I had the misfortune to be following my older brother across a ditch filled with vegetation; without knowing it he stepped on a wasp's nest and by the time the furious insects emerged, he was at the top of the bank and I was left among them as they discharged their rage. For a year or two afterwards, it seemed that every autumn, wasp headed straight for me, and I regarded them with a kind of fascinated terror, and then at the end of one summer, in my country schoolyard, a big pear tree dropped all its pears on the ground. The rotting fruit was alive with wasps, and through a long sunny week, I spent every playtime among the pears, crushing hundreds of wasps and keeping a count, which I recorded in a daily diary. The fearful child was transformed into a valiant centurion against the world of wasps, and now, 60 years later, if wasps invade my picnic, my sense is a confident delight that they will soon wish they had not. And as always with IRMs, the discharge of aggression brings its own sense of rightness – whether this is emerging from battle with wasps under a pear tree, or punching the nose of a bully in the playground.

Sometimes we need to battle and, just as we see in the triumph ceremony of the greylag goose, the battle against 'them' cements the bond between 'us'. We may think again of Robin Dunbar, and his social onion with its expanding layers of intimacy and importance. We have a deep need to belong to our own individual onion – to know who we are, who our closest friends and family are, what community we belong to and the values it holds dear; and all this is cemented by victory against a common opponent.

In the 2024 final of the English Football League Cup, the two teams facing off were Liverpool and Chelsea. Liverpool in 2024 were a great team, but in the weeks leading up to the final, they had suffered a spate of injuries. With many senior players ruled out, the team that finally took the field included a string of teenagers, suddenly promoted from the youth team to play with the men. In the closing minutes of extra time, the Liverpool captain, Virgil van Dijk, headed in the winning goal, and immediately after the match, he was interviewed at pitchside, asking what he thought of his team of boys. In the delight of victory, he spoke of pride – pride in the team he was part of, pride in the young boys who had stepped up, pride in what they had achieved. Pride in 'us' and the victory we can achieve together. It feeds on our need for battle, and it is hard to imagine life without it.

But our human aggression also brings our greatest capacity for evil. In Chapter 1, I described how, in many animals, battles are resolved without bloodshed, by a display of weaponry followed by submission of the weaker opponent. But in many animals, too, the instinct for victory is pursued to its logical conclusion, with a dead opponent who will never be back. I remember once reading an account of three cheetah brothers, spending an hour or two maiming an outsider who had erred into their territory; the story ended when one brother, finally bored, delivered a casual coup de grâce, and the horror of the story arises because we know this is just who we are ourselves.

In the flawlessly executed *Documents Concerning Rubashov the Gambler* (1996), Carl-Johan Vallgren tells the story of human evil through the eyes of Rubashov, the Russian gambler who, on New Year's Eve 1899, gambles with the devil and loses. The payment he must make is that he must accept eternal life . . . and as the twentieth century unfolds, he finds himself on the stinking battlefields of the Great War . . . accompanying Nazi exterminators . . . in a street in Budapest during the Hungarian uprising, the scene littered with corpses . . . with IRA bombers, blowing up a gaggle of nursery-school children in the street . . . in the destroyed villages of Bosnia, where 'a woman had been slit open like a slaughtered animal and a foetus lay by a mess of placenta and intestines'.[38] There is no escape – as evil follows evil, Rubashov must continue to live and watch.

Pan's aggression gives *Homo sapiens* – especially the male – the potential for unlimited, callous brutality. It is also essential – for defence, for bonding, for a sense of belonging in our own, unique community – and it has its own fierce need to be discharged, as we see in the baying of a football crowd when a goal is scored. We can neither let it free nor live without it. In the next chapter, we turn to Spock's attempts to manage this dilemma.

CHAPTER 9

The ten commandments

Adjusting to big numbers

Homo sapiens has existed for perhaps 300,000 years, preceded by several million years of evolution of other *Homo* species. Over these millions of years, our social instincts evolved in the context of small hunter-gatherer communities. All of this changed around 10,000 years ago, with the rise of agriculture and civilisation. Tribes of perhaps 150 acquaintances or fewer (remember Dunbar's number) were replaced with large, anonymous societies. As we have seen in the last three chapters, humans come into the world pre-equipped with elaborate social scripts to manage bonding, cooperation, exchange, group defence and group aggression. These scripts are not meant for societies of thousands or millions of individuals.

At the same time as group size has exploded, shared knowledge has produced a similar explosion of technology. Friendships previously cemented with a touch of the hand may now be made over thousands of miles on the tiny screen of a smartphone. Our arsenals of weaponry have the capacity to destroy not just a visible enemy, but all life on earth. Pan is subtle, sophisticated and powerful – but left on his own, Pan cannot possibly deal with this.

We have seen how, in many animals, conflicts are managed by complex systems of social signals. There is challenge and a display of weaponry, but also many inhibitory signals, preventing the stronger animal from doing serious harm to a weaker opponent. We too use complex social signals both to promote and also to inhibit our aggression but, as in any other animal, these signals need face-to-face contact. To inspire mercy we need to be there to see the smile, the bow, the look of trust or fear on a child's face. Both Lorenz and Eibl-Eibesfeldt argued that, in modern warfare, the signals are missing and the inhibition is gone. Lorenz says:

> Only thus can it be explained that perfectly good-natured men, who would not even smack a naughty child, proved to be perfectly able to release rockets or to lay carpets of incendiary bombs on sleeping cities, thereby committing hundreds and thousands of children to a horrible death in the flames.[1]

From Eibl-Eibesfeldt: 'If one asked a bomber pilot to kill his victims one by one, he would be outraged at the suggestion.'[2]

This failure of inhibition reaches its extreme in modern warfare, but Lorenz and Eibl-Eibesfeldt thought it began with the first invention of weapons. Even a hand axe, they suggested, injures far more seriously than bare hands, and instincts developed to manage the harm done with fists do not prevent a skull from being crushed by an axe. I suspect the same in the poison of much exchange on social media. Face to face, the polite residents of my sleepy English village would not dream of hurling abuse at an unhelpful neighbour, but when they sit down at their keyboards, the gloves of elementary decency come off.

In *Human Ethology*, Eibl-Eibesfeldt devotes several pages to customs that, across many tribal societies, limited permissible aggression and the scale of damage that could be done.[3] These pages make fascinating reading. Among the San Cristobal Melanesians, a man could be attacked and killed if he was on the ground – but not if he was climbing a tree or fishing. Among the Murngin (now more commonly referred to by their own term, the Yolŋu) of northern Australia, an offence between one community and another could be repaid with a highly ritualised battle. The men from the offended group could throw spears at the offenders – but the latter were allowed to run and dodge, and, among them, members of the offended group would run too, inhibiting too violent a throw from their friends. If two groups of the Tsembaga in New Guinea could not resolve their differences by argument, they cleared a fighting area for a 'small' war. Arrows might be shot, but with feathers removed so that they would not fly straight. Evidently, these customs diminishing the impact of conflict find little reflection in the conduct of our own large-scale, organised, industrialised invasion and subjugation of other nations and cultures.

In small tribal communities, Eibl-Eibesfeldt points out, it is common for men to wear elaborate ornaments of teeth and feathers, to go about their business displaying weapons and, in general, to give a demonstration of their quality as a warrior. Among friends these displays cause no harm, and may promote a sense of shared strength, but in an anonymous community, the effect is quite different. Now, a vibrant display of male strength releases aggression. As a teenager, I had a friend with a particular way of walking, suggesting a kind of cocky challenge, and it sometimes happened that strangers in the street stopped to ask if he wanted a fight. Through his teens and twenties, my younger brother played in a series of punk bands, with spiked hair and FUCK OFF written in studs across the back of his leather jacket. A gang of youths once awaited the band in the street after a gig, and once the battle was over, my mother was called out at dead of night to get the whole group released from police custody.

'In all civilizations,' writes Eibl-Eibesfeldt, 'we experience a process whereby the man becomes more drab. His dress becomes plain, the male ornamentation is reduced, his weapons are completely abandoned.'[4] 'Drab' is certainly a lovely word to describe the peaceful rows of commuters who, every weekday morning, stand patiently awaiting the train that will whisk them to their London offices. Eibl-Eibesfeldt also notes that women's decorative displays are quite different – in many cultures they are used to promote bonding, and on our streets they remain common.

Pan struggles to control aggression in an anonymous community, and his elaborate modes for managing bonding, cooperation and mutual care are equally challenged. The mechanisms we discussed in Chapters 6 and 7 – the shared laughter and dance, the shared goals, Fiske's communal sharing and one-for-one matching – all these are suitable for a group size in the region of Dunbar's number, for relations between people who all know one another. In the modern world, however, we need to live in peace with thousands or millions of unknown strangers, and to manage cooperative endeavours on an immense scale. Pan cannot do this, and thinking about the mismatch between our modern world and the world for which we evolved, it is perhaps astonishing that it can work at all. Yet it does work, not at all perfectly yet remarkably well. For this, we need the flexibility of Spock, changing morality from a local, small-scale response to a universal principle.

Big Gods

In the ten commandments of Moses, we have a simple set of ten rules that are supposed to provide the framework for moral behaviour. The idea that ten rules can capture the complexity of human behaviour is pure Spock. It is obviously too simple, but also immensely powerful. The anthropologist and psychologist Joseph Henrich argues that belief in 'Big Gods' is a recent phenomenon in human history, a belief with profound benefit for the growth of large, anonymous societies.[5]

It is perhaps surprising to us today, but the data show that, in small tribal societies, there has always been belief in supernatural agents and gods – but these gods are little concerned with human behaviour and morals. Cross-cultural research shows that belief in moralising gods increases from below 10 per cent in the smallest scale societies, to above 50 per cent in large modern states. With the growth of civilisation, Big Gods appeared at many different times, and in many parts of the world.

As examples, Henrich cites the code of Hammurabi, a Babylonian text from approaching 2000 BC, bringing the authority of the gods Marduk and Shamash to demand justice, fairness and well-being; the early Chinese Lord on High, able to intervene in human affairs and concerned with right behaviour; ancient Egyptian concern with Maat, or right behaviour; the early Greek gods, who intervened in human affairs and enforced public morality. The modern instances are familiar to everybody – Judaism, Christianity, Islam, Hinduism with its link of morality to fate in future lives – all with strict rules for the correct treatment not just of our family and friends, but of the whole, anonymous community.

Evidently, universal moral rules promote peace and harmony – especially when they are backed up by an all-knowing God who punishes transgression. They also provide an environment of trust for large-scale cooperation. Henrich points to an early Greek cult dedicated to Mercury and Hercules that, in a major maritime hub, enabled long-term trade through divine oversight of public oaths. In laboratory experiments, belief in a punishing god is linked to reduced cheating – while the reverse is true for belief in a benevolent god, and across nations, greater belief in hell is associated with lower crime rates. We hardly need laboratory experiments, however, to persuade us that the rules 'Thou shalt not kill' and 'Thou shalt not steal' discourage killing and stealing. If an all-knowing God is watching, it is a foolish person who persuades themselves that perhaps he will not see.

Big world religions use characteristic methods to build and stabilise themselves, in part based on the pre-existing innate releasing mechanisms of Pan. We have already seen how bonding is promoted by shared beliefs, rituals and music, and in big religions, these are raised to a high art, with a complex background of mythology, elaborate formalised rituals, singing and swaying together. There are picturesque leaders to be copied, marked out by swaying feathers or flowing robes. There are immense signals of authority and value, with resources beyond all reason poured into the construction of enormous cathedrals decorated with the culture's most valued art.

Henrich suggests that big religions promote themselves with CREDs – credible indications of true belief and commitment. There are fasts, painful rituals, often abstinence, and it is hard to believe that a person would starve themselves or renounce sex unless they had a real reason to do it. Henrich cites the outbursts of religious fervour that, in the early Roman Empire, were triggered by priests of the goddess Cybele castrating themselves in public, and the persuasive influence of extreme rituals devoted to Murugan, the Tamil god of war. In Christianity, we have the abstinence of Catholic priests, self-flagellations and hairshirts.

The argument, of course, is that something new was needed to allow large societies to grow – and that big religions spread precisely because they work. With a system in place to enforce large-scale morality, there is reduced social conflict and increased universal trust, providing the context for flourishing cities of strangers, extended systems of political power, the assembly and direction of huge armies. An intriguing historical study compares the lasting power of a variety of communes established in North America in the nineteenth century. Faced with the many challenges that always face group stability, religious communes lasted an average of four times longer than the non-religious.

The spread of success is a process of evolution but, again, not genetic evolution. The absence of Big Gods in early human societies shows that they are not in themselves genetically programmed. Instead, they are a result of the cultural evolution we discussed in Chapter 5 – ideas spreading based on their utility, by the imitation of people and cultures who appear successful and, often, by the tendency of a more powerful culture to impose its beliefs on weaker others. The spread of big religions such as Christianity and Islam is obvious, sometimes with beliefs actively promoted or imposed, sometimes with members of other cultures wishing to emulate the beliefs of the successful. Peace was needed, peace works and, in big religions, it could be created.

Of course, as we all know, there is a catch. Few religions promote universal peace – they promote peace within the group of believers. The first of the ten commandments is 'You shall have no other Gods before me' and, as history shows, Big Gods have promoted war between mutual unbelievers just as often as peace between believers. Spock's abstract ideas have spread across most corners of the world but, as yet, they have not been sufficiently abstract to encompass all people under a single moral code. Religious beliefs have massively enlarged the group of 'us', but only up to the defined borders of the believing group.

Henrich also asks whether big religions are the only way. Intriguingly, research suggests that religion and government are to a degree interchangeable – with increased trust in governmental institutions, there is a decrease in religious belief and in distrust of atheists. The argument is that we need *something* to support order in an anonymous community, and although big religions can fill this need, there are other social structures that can also do the job. That said, governmental institutions generally lack many of the appeals to Pan that big religions use so effectively – the rituals, the shared mythology, the singing in massive cathedrals. I should perhaps believe even Boris Johnson if, to manifest his authenticity, he were to offer himself for public castration.

The moral bottom line

Perhaps nothing illustrates the relations and tensions of Spock and Pan as vividly as big religions and their urge for a universal morality. The need for both players to explain our minds, the tension of opposing forces, the dependence of Spock on Pan and the dissatisfaction of Pan when some of his precious mental modes are denied. I sketched much of this in the Introduction. Now it is time for another look.

Big religions use the external authority of God to explain the structure of the moral code. The seventeenth-century philosopher John Locke, despite a strong belief that moral principles could be worked out from experience, drew the line at atheism. He thought that 'promises, covenants, and oaths, which are the bonds of human society, can have no hold upon an atheist. The taking away of God, though but even in thought, dissolves all.'[6] In *The Brothers Karamazov* (1880), Fyodor Dostoyevsky depicts the passionate struggle between worldly reason and immortal faith. Representing reason, there is the fearless thinker Ivan . . . representing faith, his saintly brother Alyosha. As the tortured Ivan sees, if the mystery of God cannot be accepted, then 'everything is permitted' . . . leading on, with apologies for the spoiler, to the central event in the novel. Thinkers through the ages have quailed before the moral abyss that could be left by the removal of God.

Another refuge against chaos is the belief that morality can be deduced by reason, as famously defended by Immanuel Kant in his 1785 book *Groundwork for the Metaphysics of Morals*. As Steven Pinker puts it in *The Blank Slate*,

> morality has an internal logic, and possibly even an external reality, that a community of reflective thinkers may elucidate, just as a community of mathematicians can elucidate truths about number and shape . . . The difference between a defensible moral position and an atavistic gut feeling is that with the former we can give *reasons* why our conviction is valid.[7]

I rarely disagree with Pinker, but to me it is obvious that morality cannot be a rational matter. In the Introduction, I mentioned the philosopher David Hume's argument that reason can only explain so much in our mental lives. To reason about anything we need a starting point, from which our reasoning progresses, and something outside reason is needed to provide it. We imagined the child abuser who refuses to accept any starting principle and, in general, Hume's argument is especially clear in the case of moral reasoning. In *An Enquiry Concerning Human Understanding* (1748), Hume said, 'Morality is nothing in the abstract nature of things, but is entirely relative to the sentiment or mental taste of each particular being.'[8] Lorenz loved Kant, but on this subject he puts it like this:

> It is hard to believe that a man will refrain from a certain action which natural inclination urges him to perform only because he has realized that it involves a logical contradiction. To assume this one would have to be an even more unworldly German professor and an even more ardent admirer of reason than Immanuel Kant was.[9]

As evolutionary psychologists have often argued, the starting point, the thing that we need to underpin our moral code, is not God, and it is not reason. Against thinkers for millennia, our moral core is our own animal side, our own Mr Hyde – the Pan and his values that we are built with. We protect our children, we value fairness, we do not steal from our friends, we fight for our community and if necessary lay down our lives for it simply because this is the way we are. A herring gull chick pecks at the orange dot on its parents' bill, and we pursue our moral imperatives, with all their exotic cultural elaborations, for just the same reason.

In the Introduction, along with Hume I also quoted from Darwin, and his view that advanced human morality derives originally from social instincts in part shared with other animals, refined through human evolution by increased understanding and sympathy. Darwin saw all the parts of this puzzle – the role of instinct, the expanding contribution of reason, the wish to avoid 'reprobation of the one God or gods, in whom according to his knowledge or superstition he may believe', the gradual extension of sympathy beyond family to all humanity and even to inanimate objects.[10] He appeared to understand, too, that the buck stops with us:

as the feelings of love and sympathy and the power of self-command become strengthened by habit, and as the power of reasoning becomes clearer so that man can appreciate the justice of the judgments of his fellow-men, he will feel himself impelled, independently of any pleasure or pain felt at the moment, to certain lines of conduct. He may then say, I am the supreme judge of my own conduct.[11]

By 'right' and 'wrong', we refer not to something external but to something in ourselves – the moral dictates of our own innate dispositions and their cultural elaborations.

As I have said, it might seem quite extraordinary that big societies and big religions can work at all. In my view, they have worked so well exactly because they build on the innate blueprints that Pan provides. In the form of these blueprints, we have IRMs for managing relations within a community, including all the modes suggested by Fiske and Haidt. They manage care, cooperation, authority, loyalty. When it comes to relations with outsiders, we also have IRMs for attack and community defence.

In all aspects of our social lives, these IRMs compete for control, and in big religions, Spock just adjusts the weights. The admonition to turn the other cheek is not an intellectual bolt from the blue. It is Spock's encouragement to favour the mental mode of acceptance over the mental mode of battle, and it works because the acceptance mode is already there – exactly in line with Eibl-Eibesfeldt's account of a friend cutting him off on the road, or with a father who smiles as a one-year-old grabs a stick and hits him over the head. Religions do not really introduce a new moral code. They suggest a different weighting between the codes we already have.

The rules are too simple

Perhaps nobody, or at least almost nobody, really believes 'Thou shalt not kill'. As I said in the Introduction, Pan quite clearly tells us that the real meaning is, '. . . except sometimes.' The abstract rules of Spock capture things that are important, even essential, but in any real situation, many other things matter too. You kill when your husband's crazed lover springs up yet again from the bathwater, and watching *Fatal Attraction*, the entire audience thinks . . . 'At last!' You prepare to kill when your god tells you that you must sacrifice your only son. A simple rule will never be the whole story, because our lives, like the lives of all animals, are filled with conflicting considerations.

In my teens, I watched Akira Kurosawa's 1951 film version of *The Idiot*, masterfully realised in the world of post-war Japan, and overcome with admiration and the sense of its rightness, I turned to Dostoyevsky's Russian original novel, published in 1869. For the next 30 or 40 years, if I was asked to name my favourite novel, it would always be *The Idiot*. In a quite flawless way, Dostoyevsky chronicles the life of a truly good, Christ-like man – burned up with sympathy for the flawed, unable to harm, unable to think or speak ill. I felt that it was an essential guide to the life one would wish to live (though even in my teens, when I wanted to be a novelist, I embarked on my own novel explaining that *The Idiot* is only half the story, and that sometimes we need not only the flawless Prince Mishkin but also the all-too-flawed Rogozhin). Somewhere around age 60, I dug out *The Idiot* and read it again, and, of course, I discovered that Dostoyevsky was way ahead of my innocent teenage self. Prince Mishkin is a perfectly good man, who answers Dostoyevsky's own longing for Christ, but he is also doomed. Faced with many-sided reality, the abstract ideal can never be enough.

We all know, too, how an abstract moral rule, once it is put into place, can generalise far beyond what is reasonable or balanced. We have the sterility of no sugar in the porridge, no music on a Sunday (or perhaps any other day), no flirting with boys on the school bus, no reading the Devil's works. It is in the nature of abstraction that it discards everything outside itself – the joy of the music and its essential bonding power, the bubbling need to laugh with a friend, sometimes the need to raise a fist instead of turning the other cheek.

Simple rules can be sterile, and they can also be immensely dangerous. Throughout history and across the world, the atrocities of the Crusades or the Spanish Inquisition are committed in large part because the rules of the religion are believed. Spock's moral rules are often great – but they should never actually be believed.

Disagreements

Cultural values evolve, and often it is tempting to believe that there is progress in some absolute sense. When we think of imperialism and slavery, it seems clear that the views of today are more 'right' than the views of yesterday. Again, with the development of modern thinking, there has been an adjustment of weights. Now, modes of equality and fairness are extended outside our cultural and national borders, to include the rights of everybody. Modes of aggressive exploitation are down-weighted – though, of course, I would never for a moment hope that they are gone.

I have no problem with the claim that my own view of slavery is more 'right' than the views of an eighteenth-century slave captain, but I would never say that, in any objective sense, it is more 'true'. When I say what I hold to be right, I mean exactly that – the standard of what is 'right' is my own set of values, the product of my own, relatively universal IRMs and my own, idiosyncratic cultural background. Faced with an eighteenth-century slaver, I think it would be impossible to prove to him that he is mistaken. What is possible is to tell him that I will not tolerate his behaviour and, if necessary, to fight it.

Weights vary not just between cultures but between individuals. So too do the specific details attached to each of our innate moral modes. For one culture or one person, the symbol of the cross has enormous value while, for another, the same is true of a mother-of-pearl necklace and armband. In Chapter 7, I discussed Jonathan Haidt's proposal that, in large part, the ferocious disagreements of American Democrats and Republicans can be traced back to the relative weights given to moral modes – care/harm, loyalty/betrayal, sanctity/subversion and the rest. Steven Pinker points out the tension that arises when one person's moral code is 'amoralised' by another – homosexuality and drug use can be a taboo for one person but a free lifestyle choice for another.[12]

The sense that one's own beliefs are in some sense 'the truth' is overwhelming. Perhaps every devotee of a big or small religion believes that their particular belief is right and that all the others are wrong, with the sense that the meaning of 'right' and 'wrong' is something bigger than the particular culture they happen to have adopted. It does not matter that just a few moments' reflection shows how very unlikely this is. Once the weightings and the cultural details are in place, they have enormous force in shaping our feelings, our lives and our sense of stability in the world. A young friend once told my son that she believed in God 'because it would be so depressing not to'.

New moral rules arise at the lightning speed of cultural evolution, often within a single lifetime, while the needs and promptings of Pan remain unchanged. As we focus on one new abstraction, it is inevitable that we feel losses as well as gains. In the Introduction, I used current battles over political correctness to make the point. With modern political correctness, the value of equality is radically upweighted. There is a push, in itself heart-burstingly positive, to treat all people the same – Black, white, straight, gay, fat, thin. It is easy to see how well this fits the needs of a truly global community. It is easy to feel the welling sense of rightness that might follow making all people 'us' – as Darwin foreshadowed a century and a half ago.

Meanwhile, the needs and sometimes merits of the 'them' mode are lost. There is much current discussion over the thesis that 'wokeism kills comedy', with professional comedians vigorously putting both sides of the case. There are many sides to comedy, so comedy is certainly possible without discrimination . . . but on the other side, I can hardly doubt that the tension between 'us' and 'them' is often funny. A joke that begins 'Two Welshmen go into a bar . . . ' sounds far more promising than 'Two humans go into a bar . . . ' – unless, perhaps, the intention is to mock a woke comedian. Pan has a strong mode of 'laughing at', and the two Welshmen joke prepares to exploit the bonding power of our (in this case, English) union against 'them', with all its accompanying sense of our own, unique community. It is not subtle and it is not clever – it is no more than the greylag triumph ceremony – but it is funny and it does reflect something firmly built into us. We may very well wish to give it up for the sake of embracing a wider humanity, but what we have given up, in part, is something we used to tell us who we are.

Spock's rules tend to over-generalise and, arguably, this can happen with the upweighted equality rule. A few years ago, I was discussing one colleague with another. She was not sure who I meant, and I said, 'He's the Black Frenchman.' Looking at me a little sideways, she told me that nobody else in our institution would have said that – though 'Black Frenchman', in the context of our very heavily white workplace of many nationalities, was certainly the best possible way to identify the person in question. Is it going too far even to mention a person's identifying characteristics? It is easy to understand where this idea comes from, but not easy to decide exactly how far one should take it.

Built so deeply into our Pan, the 'them' mode suppressed in one place simply pops up in another. The person who believes in themselves as a well-intentioned liberal, opposed to any form of discrimination, can reaffirm this identity by finding a new out group to attack. We have 'the only people I hate are the haters', or perhaps shots taken at older white males. Embracing all of humanity does not come easily. We have upweighted equality, but then find we are using the 'us' and 'them' modes to defend it.

According to the ideas of Lorenz, there is only so much that Spock can do to manage such conflicts. It is not just that each social IRM has its own, important role in managing our lives. Neither is it just that discharging them feels good, though as we prepare to laugh at those innocent Welshmen, it certainly does. It is that each one has its own spontaneous need to be discharged, building up until it can be released, meaning that simply bottling it down is ultimately fruitless and potentially dangerous. As I said in the Introduction, it cannot work for Spock simply to tell Pan to be quiet.

Much more promising, as Lorenz argued, is to create harmless opportunities for discharge, and modern culture finds many ways to do this. Enormous bodies of people worldwide sit before their Xbox spellbound in team war games. Enormous bodies of fans cram into football stadiums, screaming that the other team are

'Manc scum' and heartbroken when the Manc scum do something idiotically trivial – kicking a big ball into a net. There can be real violence after the match but at the same time, I suspect, the screaming does much to discharge and satisfy the urgent needs of the 'them' mode. Standing up from the Xbox, the merciless killer is ready once more to cook the children's dinner and put them to bed.

It's complicated

Overwhelmingly, the complexity of 'morality' shows the conflicted, discordant world of Spock and Pan. The IRMs of Pan provide blueprints for social choices – often conflicting, each with their own weight as we choose whether to keep or share, to laugh or cry, to choose battle or peace. This is already complex, but with his infinite ability to generate new ideas, Spock adds infinite further complexity, adding rich cultural details to our broad moral modes, and often, with his ability to focus, throwing his weight behind one choice or another. Many of his ideas are brilliant, allowing civilisation to function in a world quite different from Pan's. Many others are arbitrary, over-generalised and harsh.

Even a single mind struggles with its conflicting voices, and in large-scale human societies, millions of different minds need somehow to live together. Religions and moral systems attract us with their simple rules for right behaviour – but in the dialogue of Spock and Pan, moral questions can never be simple.

CHAPTER 10

Keeping things together (2)

The law

In Chapter 7, we considered the elaborate innate releasing mechanisms that guide smooth human interactions, bringing the desires to give as well as take, to ensure that exchanges are fair, to defer to accepted authority, to protect the sanctity of the group and its practices. When these mental guidelines are violated, we have specific mechanisms for correction – we regard the offender as 'blameworthy', we gossip about their failures to other members of the group, we may exclude, criticise or otherwise punish. All of this, doubtless, is based heavily on Pan – a set of innate mechanisms allowing small groups of early humans to function.

In Chapter 9, we considered how moral rules have been generalised to bring stability to the large, anonymous societies of the modern world. Now, we have the universal morality of big religions, with their accompanying large-scale rules for moral censure and divine punishment.

To keep things together, however, universal moral rules have not been enough. Accompanying these we have the law, with its astonishingly detailed prescription of what is and is not allowed, and elaborate mechanisms for preventing violations. We have formalised ideas of justice, rights, responsibility, blame. We have the uneasy sense that the law is indispensable but an ass. As usual, I think that none of this makes sense without the dialogue of Pan and Spock. To understand it, we need our biology, not just our simplified and often misleading ideas.

Again we begin from Pan

Throughout history, thinkers from Aristotle to today's legal scholars have sought a basis for the law. As we saw for moral rules, the two big candidates have been gods and reason. Either the laws are set down by an external divine being, or they can be deduced from some coherent, abstract principle. Often the two have come together in the idea of 'natural law', embraced by the Greek Stoics, by Roman lawyers such as Cicero, by Thomas Aquinas and the Catholic Church, and still with appeal today. The core idea is that law exists in nature, often because it was put there by God, and that as rational beings we can deduce it. This quote from Cicero captures it nicely:

> True law is right reason in agreement with Nature; it is of universal application, unchanging and everlasting . . . It is a sin to try to alter this law, nor is it allowable to attempt to repeal any part of it, and it is impossible to abolish it entirely . . . [God] is the author of this law, its promulgator, and its enforcing judge.[1]

Or, as I previously quoted from the American Declaration of Independence:

> We hold these truths to be self-evident, that all men are created equal, that they are endowed by their Creator with certain unalienable rights.

As with morality, it is reassuring to think that 'true' law is out there and, with reason, we can work out what it is. Certainly there can be remarkable constancy, across cultures and across millennia, in judgements of what is wrong, how wrong it is and how much punishment should be handed out. In a fascinating study, the psychologists Daniel Sznycer and Carlton Patrick asked whether sets of ancient laws would feel right to modern experimental participants.[2] The laws were drawn from two sources – the Laws of Eshnunna, written down in Mesopotamia around 1770 BC, and the Chinese Tang code from around AD 635. These written codes were useful because they specified offences and the prescribed punishments; Sznycer and Patrick gave their participants the offences, and asked them to estimate how wrong they were. The offences were often exotic, and quite outside the experience of the modern raters – perhaps deflowering the slave of another person, or allowing a fellow citizen to be gored by a dangerous ox. The original laws specified the punishments, in terms of years in prison (Eshnunna) or the amount of a fine (Tang). To a remarkable extent, the judgements of modern people matched the punishments originally specified – the more wrong the action seemed today, the greater the punishment specified in the ancient civilisation.

There is consistency here but, as with morality, the search for natural law also runs quickly into difficulties. For Thomas Aquinas in the thirteenth century, violations of natural law included killing the innocent, blasphemy, adultery and sodomy. For Western culture today, some of these remain and some do not, and it is hard to imagine any rational process that could determine whether Thomas Aquinas was right, and sodomy is indeed against natural law, or whether I am right, and sodomy is a personal choice.

Beyond natural law, there have been many brilliant attempts to put law on some rational, coherent basis. At the heart of this, there is often the concept of justice – that people should be treated fairly and equally. In Roman law as codified in the *Corpus Juris Civilis*, justice was defined as 'the constant and perpetual wish to give everyone that which they deserve'. In the twentieth century, the legal scholar John Rawls put forward a fundamental theory of justice.[3] In an elaborate thought experiment, he imagines that everybody first agrees to laws without knowing their own position with respect to the principles that will be laid down. They do not know whether they will be rich or poor, healthy or ill, talented or mentally disabled, and so on and so forth. Accordingly, laws are agreed that should be fairest to everybody, no matter their position in society. They are the laws that 'free and rational people' would choose to ensure that, should they turn out to be the vulnerable in society, they come to the least harm.

Despite my admiration for the brilliance – and indeed the utility – of this sort of endeavour, of course I do not agree with it. Not surprisingly, I think instead that the law, like the moral rules of big religions, has its origins in biology, and our innate sense of what should and should not be allowed – exactly the argument, in fact, that Sznycer and Patrick, in agreement with other evolutionary psychologists, use to explain their findings.

It is Hume's argument again – the law needs a starting point, and the starting point is us. We are born to value fairness, respect for one another's rights, the sanctity of our practices, and so on, and to seek a consistent or rational basis for these things is hopeless. The relative weight we put on our different values varies from person to person, from culture to culture, from moment to moment. To hope for a coherent, principled basis for all this is no more sensible than hoping for a coherent, principled account of any other animal's behaviour – for the turkey mother who ferociously protects her young until it fails to make the right sound, or the territorial fish who decides whether to accept or drive off a potential mate. We do not know whether we 'should' be for or against sodomy because there is no universal way that we could know that. As Jonathan Haidt argues, if we are under the influence of the urge to fairness, we may conclude that this is a personal choice.[4] If we are under the influence of our urge to sanctity, and enforcing common behaviour across the group, we go with Thomas Aquinas. The basis for the law is not an abstract matter for legal theorists. It is the biological mess that is us.

In a related vein, much debate in jurisprudence asks what is the 'proper' domain of the law. In principle, for example, should the law even concern itself with matters of private morality? In my view, again, such questions cannot be settled 'in principle'. To know what rules we wish to implement, we need all the complexity, inconsistency and conflicts of our own human nature. Of course, this means that the law indeed has a 'natural' basis. Rather concretely, though, the 'nature' concerned is human nature. If we firmly accept this, then some of our most entrenched and precious thoughts have to change. Laws can be justified, but not in the way that we usually think . . .

Human rights

Let's begin with 'rights'. What is a human right? Passionate debates focus on the right to work, to vote, to terminate a pregnancy, to be safe from violence or theft. Rights are often felt to be 'real' – universal, absolute, inalienable. At the same time, there is concern over what criterion could be used to decide if a right exists. As the legal scholar James Griffin puts it: 'The term "human right" is nearly criterionless. There are unusually few criteria for determining when the term is used correctly and when incorrectly – not just among politicians, but among philosophers, political theorists, and jurisprudents as well.'[5]

If there is no external agency determining what the law 'should' be – if it is all relative to human nature and human desires – what do rights mean? What is the correct 'criterion' and the correct usage?

To me the answer is obvious, even if it feels wrong. A 'human right' is simply an agreement, no more and no less. When we say that a woman has the right to decide on her own pregnancy, we mean only that we agree that this is what the law should allow. People believe passionately in, struggle for, and may live and die by their rights. This all makes perfect sense, but not because the 'right' has any existence outside themselves. They believe and struggle because it is their choice of how they would like society to be run.

I strongly suspect that there are IRMs for creating the passionate sense of a human right, just as there are IRMs for learning a language. To some extent, also, the content of these learned rights will be consistent, reflecting the common, instinctive human values of fairness, freedom, and so on. Probably, this passionate sense of a right is important in ensuring consistency of behaviour in the members of a culture and, certainly, we use it strongly in debates over how we should and should not behave. But as we have seen, people have many instinctive forces in their behaviour, sometimes in conflict, and once we combine this with Spock's ability to fill an IRM for 'rights' with different cultural details, it becomes obvious that different cultures will often believe in different rights. One person fights passionately for his right to keep slaves, perhaps under the influence of IRMs for possession and (his own!) freedom to further his own lot. Another, equally passionately, favours the right of justice for all, and finds slavery abhorrent.

None of this is puzzling if rights are simply cultural agreements. Of course, different groups of people may come to different agreements! The First Amendment of the American Constitution prohibits the law from interfering in free speech. Free speech is a precious right to Americans, but not in other countries such as the UK. Indeed, my personal opinion is that an absolute right to free speech is a piece of ill thought-out idiocy – speech can harm just as surely as violence, and to me it seems absurd to value freedom so highly that all such harms are permitted. But when I say this, I am not saying that free speech should not be a 'right'. I am saying that, if you asked me to agree to it, I would not. It is a right in the US simply because it is accepted, and it is not a right in the UK simply because it is not.

Perhaps this seems unsettling, since we believe in our rights so strongly and use them so vitally in managing our society. We do need agreements, we need to live by them, and very often, they will incorporate strong elements of universal human values. The passionate sense of human rights promotes all of this, even though the particular rights we live by are just our own cultural agreements. Pan provides the passion in our rights, even while Spock may know they have no external, absolute validity.

Responsibility and free will

The law depends heavily on our ideas of responsibility and free will. In my opinion, once again, Spock gets us into trouble by abstracting these ideas, and trying to find an impossible basis to support them. Again, I think it is all fairly simple and obvious, as long as you approach these ideas as concrete biology rather than abstract philosophy.

Certainly we have a strong intuitive sense of our free will, and of responsibility for our actions. But what does this sense depend on? Once we abstract the idea of free will, it seems we wish to claim that our choices are free of *everything* – but is this really what we feel? We know that everything has a complex chain of causes, going back as far as we like into history. A pilot is in a hurry and lands his plane too fast, missing the tarmac and causing a minor incident. What causes the accident? The pilot made a misjudgement, and might be held to account. The company's schedule put the pilot under pressure, and might be held to account. The chairman of the company was a driven business-man, and his parents might be held to account. There is no logical way of stopping this travel back along the chain of causes. There is only the convenience of where we may stop to keep down the rate of accidents.

Surely we do not believe that our choices are free of all their antecedent causes, including all those causes that went to make up the person we are – our training, our childhood upbringing, our genetic constitution, the evolutionary forces that shaped that. If we take free will to be freedom from *everything*, we are soon in an incoherent mess.

Instead, there is another route to understanding responsibility, where all these puzzles disappear. Our culture lives by a body of agreements. We hold people responsible for honour-ing those agreements and, if they do not, we sanction them. We make exceptions where there is obvious reason for the agreement not to be met – if the person is a one-year-old child, or is suffer-ing a psychotic breakdown, or is under the influence of extreme emotion or duress. When we say a choice was free, we simply mean that none of these factors was in operation. We mean that, under these particular circumstances, a member of our culture could reasonably have been expected to follow the rules and, if they did not, they are responsible and can be punished. We are not making any extreme claim about chains of causality. Again, we are using Pan and his commitment to social order, and Pan really does not care about long chains of causality.

In my opinion, if we really believed in the abstract ideas of free will and responsibility, the law could not function at all. Of course, it is always true that causal influences on a person's behaviour can be traced back to their early experience, to their genes, to the forces of human, mammalian, vertebrate evolution. If these forces are taken to mean that the person was not free and not responsible, then, without doubt, nobody is responsible for anything, and all crimes should go unpunished. The law can only work if, explicitly or implicitly, only some causes count in removing responsibility. It is up to us which causes we include – does a history of childhood abuse mean that the rapist should go unpunished, or do we recognise the cause but still choose to incarcerate? To make the law work, I think we have to realise that such choices are somewhat arbitrary and cannot be informed by our over-simplified ideas of free will and responsibility. Other choices might also be defensible but, as a society, we just have to buckle down and draw some lines.

Once again, these conclusions can be unsettling and, indeed, serious thinkers are so unsettled that they conclude the very ideas of evolutionary psychology must be wrong. Here is a nice, confused quote from the evolutionary biologist Stephen J. Gould, which I got from Steven Pinker's *The Blank Slate*:

Perhaps the most popular of all explanations for our genocidal capacity cites evolutionary biology as an unfortunate source – and as an ultimate escape from full moral responsibility . . .

223

Chimpanzees, our closest relatives, will band together and systematically kill the members of adjacent groups. Perhaps we are programmed to act in such a manner as well ... but we cannot be blamed for these moral failings. Our accursed genes have made us creatures of the night.[6]

Gould wished to argue that, since we *can* be blamed, the causes of our aggression cannot be traced to our genes. The problem, however, is not with the evolutionary biology, it is with the wrong idea of blame. Undoubtedly, our genes belong in the causal path that led to what we are. But blame does not presuppose that there are no reasons for who we are. It simply means that, even in the absence of a modest, agreed set of extenuating circumstances, we did not obey the rules – and are now being held to account for that.

The same thing applies the other way around. When we say a person 'was not to blame' – it was the parents, or the social deprivation, or the genes – we mean literally and simply that we are not going to blame them! When we say this, we focus on the long causal chain, turning attention away from the person this chain has produced. It is the same criminal either way – but this time, we are not going to blame him.

There is no rational system

If the law is simply a set of procedures for making society function, could it do better by discarding Pan altogether? Many thinkers have imagined that we can do better than follow our animal selves. This is Aristotle and his idea of law as 'reason free from passion'. In *Leviathan*, the seventeenth-century philosopher Thomas Hobbes imagined that, without an all-powerful state to enforce the rule of reason, human life would be a disorderly, animal 'prey to passions'. As an extreme example, Eibl-Eibesfeldt quotes the Canadian psychiatrist George Brock Chisholm:

> The reinterpretation and eventually eradication of the concept of right and wrong which has been the basis of child training, the substitution of intelligent and rational thinking for faith in the certainties of the old people, these are the belated objectives of practically all effective psychotherapy . . . The suggestion that we should stop teaching children moralities and rights and wrongs and instead protect their original intellectual integrity has of course to be met by an outcry of heretic or iconoclast, such as was raised against Galileo for finding another planet . . . If the race is to be freed from its crippling burden of good and evil, it must be psychiatrists who take the original responsibility . . . With the other human sciences psychiatry must now decide what is to be the immediate future of the human race.[7]

Eibl-Eibesfeldt was a rather serious German scientist, not usually given to levity, but even he could not resist commenting, 'One cannot exactly say that Chisholm is troubled with modesty.'

In the law, the most influential version of this thinking is the utilitarianism of Jeremy Bentham and his followers. Writing in the eighteenth century, Bentham proposed that the correct basis for law is the promotion of happiness and avoidance of pain. The right legal decision maximises the overall happiness of the population, and minimises overall pain. Bentham's utilitarianism was followed and modified by many later scholars, prominently including John Stuart Mill. There is an obvious appeal to the idea that the law should be 'what is best for everybody', and if this dispenses with intuitions over what is right and wrong, perhaps indeed the law would be better without them.

But can it work? Subsequent legal scholars have pointed to many cases where this simple standard of 'net happiness' seems completely wrong. On the basis of Bentham's proposal, for example, a judge might compel a richer person to give money to a poorer one, simply on the grounds that the rich person will feel little pain over the loss, while the poorer one will feel much pleasure at the gain. We may well encourage charity, but do we really want the law to compel it?

At the heart of such difficulties is the indispensability of Pan. It is simply not true that there is a single scale of 'pleasure versus pain' on to which we can map our many human values. We have seen many of these values in previous chapters – the urge sometimes to give to our children, the urge sometimes to treat others in strict fairness, the urges both sometimes to defer and sometimes to insist, the urge sometimes to turn the other cheek and sometimes to destroy an enemy. These IRMs – and all Spock's elaborations in each particular culture – provide what it is we value and need, and they are just different things, with different weights, all varying across person, place and time. One human IRM dictates that people should help others. Another dictates the right to keep one's possessions. To use a single scale of pleasure and pain, we should need to know how much of the 'giving' urge corresponds to how much of the 'keeping' urge, and there is simply no way to settle such questions in any general way.

When the imaginary judge insists that money be transferred from the rich person to the poor, we can see that the poor person gains, but we also sense something unfair. It is Hume again. There simply is no 'rational' way to decide what is best in all respects. No matter how carefully it is designed, the best that the law can be is the usual set of uneasy compromises between the many values of Pan and the rapidly evolving ideas of Spock.

To illustrate and analyse such compromises, to show that there is no single 'best' choice, neuroethicists like Joshua Greene have used a classic moral dilemma, much beloved of moral philosophers – the trolley bus problem.[8] A trolley bus is speeding towards a fork on the track. You are standing next to the lever that determines which way it will go. At present, the lever is set for line A, and on line A, five people are standing. They will be flattened. On line B, one person is standing. Will you hit the lever and kill this one person to save the other five? In Greene's studies, people lie in a brain scanner deciding what they would do as the trolley races to the choice point.

The decisions are interesting, and so is the brain activity. First, perhaps not surprising but very important – people differ. For Jeremy Bentham, this dilemma is scarcely a dilemma at all; it is obvious which choice leads to less net pain. In agreement with that, people can see that the world is better off with only one person dead, and pull the lever. In the face of this 'rational' choice, however, there are also many people who feel that, no matter what, it is not their business to kill an innocent person. People in this group stand back and let the tragedy unfold.

Second – it all depends on the circumstances, in the classic fashion of Pan. Imagine a new scenario – now there is only line A, with five people awaiting their doom, but you have the option of pushing – physically pushing – another person on to the tracks to derail the trolley. Now, far fewer people are happy to act.

Meanwhile, what is happening to brain activity? As people contemplate the 'rational' solution, pulling the lever to save (on balance) four lives, the brain activity is complex – but a part is reminiscent of Spock and the 'multiple-demand' or MD pattern that we saw in Chapter 4, including activity in the brain's frontal lobes. When people think instead of pushing that innocent person to their death, again we see something complex – but now there is more Pan, with activity in one of those core, ancient structures we considered in Chapter 4, the amygdala.

In the law, we can write immensely detailed specifications of how our society is to be conducted – but we cannot possibly escape from the conflict and complexity of ourselves. It will be a long time, I suspect, before society sends a person to jail for refusing to push that hapless single victim on to the track.

Blame and deterrence

Perhaps the deepest conflict between Pan and Spock arises over the issue of deterrence. When somebody wrongs us, we have the strongest imaginable urge for blame and revenge. The evolutionary forces underlying this IRM are not hard to work out. In a competitive world, an animal or group that does not fight back is likely soon to go under and, in our own mental world, we know that an attack on others brings a corresponding risk of reprisal. In the law, perhaps the strongest motivation for punishment is deterrence – we believe that the risk of punishment must act to discourage the criminal act. This belief feeds directly on Pan, and the urge to pay back, and very likely, it must often work for the same reason it worked during human evolution. The potential criminal considers the possible consequences, and thinks twice.

At the same time, we know full well that punishment may not be the best way to reduce crime and that, often, it could be more effective to understand the root causes, and to change them. This alternative perspective tends to gain force with reasoned analysis. In academic criminology, for example, there is reluctance to 'stigmatise' the criminal, and an effort instead to understand what forces led to their choices. The conflict between blame and understanding plays out in longstanding political clashes. On the one side, perhaps, 'rehabilitation', and on the other, the accusation that this is 'soft on crime'. No matter how strong the blame/revenge IRM, however, I imagine that few people seriously doubt that crime has causes and that these causes could in principle be addressed – perhaps (at huge cost) by eliminating social deprivation, or by legalising the sale of drugs. Blame and revenge are powerful forces but, as always, there is more to a situation than the IRM it triggers. The person who commits a crime, like the law that judges them, is under the control of many competing influences, including the many sides of their own nature and the many facets of the world they live in.

Though I doubt that I am hopelessly 'soft on crime', I do think we should never rush to conclude that the person who wronged us is entirely different from ourselves. They are a complex mixture – we are a complex mixture – and though the mixtures differ, creating each unique individual, it may be no bad thing to recognise the many common ingredients.

Meanwhile, why am I not softer on crime than I should be? The problem is that, in our staggeringly complex world, we rarely understand things so well that we can be certain how things will work out. I imagine that the principle of deterrence does reduce some crimes. I imagine that opportunities for rehabilitation and re-entry into society do reduce some crimes. Legalising drugs would, of course, remove at a stroke one of the largest pillars of organised crime worldwide – but quite possibly with other, increased human costs. We argue about the merits of these things in part because of our own, conflicted nature, but in large part, too, because we just don't know the answers. Discouraging Pan, with his passion for revenge, is particularly difficult when sometimes he may actually be right.

Simple ideas may not work out

In a paper called 'Evolutionary psychology, moral heuristics, and the law', the evolutionary psychologists Leda Cosmides and John Tooby discuss how social decisions can go wrong when they are based on too-simplified abstractions of our instinctive human values.[9] Their examples show how the real world is more complex than our apparently well-reasoned ideas.

One of these examples concerns the apparently simple case of rent control. When a person is poor and homeless, this is a strong stimulus for Fiske's communal sharing mode. We want to pitch in and help, especially if we can see that this poverty was partly a matter of bad luck, with the sense of 'there but for fortune . . . ' A natural thought is that homelessness might be reduced by regulating rents, providing more low-cost accommodation, but Cosmides and Tooby cite a comparison of US cities that shows this just doesn't work. Across cities, homelessness was not predicted by rates of unemployment or poverty. Instead, it was predicted by rent control laws – the cities with rent control had *more* homelessness. The true situation is more complex than our communal sharing mode captures. The strong effect of rent control, as it turned out in this culture, was to reduce the amount of rental accommodation put on the market.

Even more striking is Cosmides and Tooby's analysis of Marxism. Again, under the influence of the communal sharing mode, we might opt for a social structure where everything is shared – resources, labour and the fruits of that labour. The appeal is obvious, but so is the result, with few communist economies leading to flourishing social well-being.

Cosmides and Tooby argue that the root cause comes from another aspect of our human nature. When a group comes

together to share, they are driven no doubt by the delight of
the sharing IRM, but also by the reluctance to be judged as a
non-contributor. This is the evolutionary point we considered
earlier – for sharing to evolve in a stable way, our IRMs need
to enforce that we share *specifically with those who will also share
with us* – and that we will drum out free riders who take more
than they give. However, as a part of our human Pan, IRMs to
criticise others for their social failings, and to avoid the criticism of
others, evolved for small human groups. Criticism works well when
each person in a group knows the others, and social judgements can
easily be formed, expressed and acted upon. It is not really adapted
to large, anonymous societies, and research shows that, when a
group can work together to a common goal, average contributions
start to fall as group size increases. The situation is made worse by
a typically human positive-feedback loop – as a person perceives
that others are not doing their share, they too are not so willing
to contribute, and through this so-called ratchet effect, the entire
structure can crumble. Here is an example from Cosmides and
Tooby's paper; when I read this, I just wrote 'Wow' in the margin:

> One of many natural experiments was provided by agricul-
> tural policy in the former Soviet Union. The state nationalized
> farmland and forced farmers to organize their labor as a col-
> lective action. But they allowed 3% of the land on collective
> farms to be held privately, so local farming families could
> grow food for their own consumption and privately sell any
> excess. The results were striking. Estimates at the time were
> that this 3% of land produced 45% to 75% of all the vegeta-
> bles, meat, milk, eggs, and potatoes consumed in the Soviet
> Union . . . The quality of land on the collectively-held plots
> was the same; their low productivity was due to the itera-
> tive ratchet effect. People shifted their efforts away from the

collective to the private plots. Without these private plots, it is likely that the people of the Soviet Union would have starved. In China, when all peasant land was collectivized into mass communes of roughly 25,000 people apiece, the result was the largest famine in human history.[10]

These examples sound a loud bell of caution over the attempt to replace our complex body of social IRMs with simplified legal structures. With the best intentions in the world, Spock steps in and makes a mess of things. In *The Blank Slate*, Steven Pinker argues that we should not use our instinctive gut reactions as a moral guide. He says that things should not be banned because they fail the 'shudder test', but because we have good *reasons* to ban them. I think, on the other hand, that we cannot easily do without the shudder test. It is not only that we need Pan to give reasoning its starting point (Hume again), and to define what reasons actually count as 'good'. Even if we have decided what we want to achieve, with millions of years behind it, the shudder test can still be a valuable guide in predicting whether our plans are likely to work out.

Reason and passion

In the law, as in private morality, it is the usual rich dialogue of Pan and Spock and, very often, what we think about this dialogue does not really make sense. Our judgements of others, and our decisions how to treat them, rest heavily on 'holding them responsible' – but this does not really depend on an abstract idea of free will. 'Holding somebody responsible', I think, is just another of our IRMs, allowing us to support them when their behaviour matches our rules, and to criticise or punish when it does not. Along similar lines, I suspect we have a strong IRM to believe passionately in those 'human rights' that we hold dear – and this works well even though rights are not inalienable, do not have an objective reality, and accordingly can vary from person to person, from culture to culture, and even from one moment to another in our own lives.

We have an urge, extending over millennia, to find an objective or rational basis for our laws, but when it comes to human choices about how society should be run, objectivity and rationality can never be enough. As I said in the Introduction, when it comes to reason free from passion, Reese Witherspoon was right and Aristotle was wrong. The law is not reason free from passion – it is Pan and Spock again. Pan gives us our many, often conflicting values and moral imperatives. In elaborate detail, Spock works out how these values can be implemented in a large, complex, civilised world.

PART III

Ambition

CHAPTER II

We don't want to be rich and happy

Rising to the challenge

A few years ago, my wife and I were visited by very old friends from America. We spent a few days showing them around the tiny country lanes, green fields, windswept trees and spectacular coast of north Devon, still the home of most of my family. One afternoon, the typical Devon mixture of sun breaking through rain, we were at a beach, with rocks, sand, waves and the usual towering cliff right above the surf. Deciding to climb the coast path – a stiff challenge in Devon, where the path is always going either steeply up or steeply down – I made it to the clifftop and looked down at the specks of my friends below. In the surf where the waves were breaking, a dog was running wildly up and down in a vision of pure joy. The dog, in the way of dogs, was going nowhere – it was just running because running through the surf is what a dog loves to do.

The scene tells us about dogs and the burning innate releasing mechanism to run that is there in a pack animal that is born to chase down prey. But the scene also tells us about the man who looks down on that dog from the clifftop. There is no sensible reason to slog to the top of a cliff when the only thing that will happen is that you have to slog down again. (In wet weather, down can be even harder than up on the slick Devonshire coast path.) The man slogs, as the dog runs, because he is born to it, and there is surely something transparently symbolic about getting to the top.

My mother loved life, and certainly I deeply loved my mother. On her eightieth birthday, we had music, dancing and camping in the field behind the farmhouse, and though dancing is not easy on the grass, my mother and sister were jiving as they always did. A few years later, going up the stairs, something happened in my mother's knee and, though it was never explained, that was the end of her dancing. I was in the habit of phoning her on Sunday evenings, and now I would always be asking about her knee, and her life now pretty much confined to the house. Often she sounded a bit down – which, for my mother, was a massive change.

One winter Sunday, she picked up the phone and I could immediately tell she was in a barnstormingly good mood. It had been snowing. With her walking sticks and half-crippled knee, she should have been wrapped up by the fire, but instead she had decided to take the winter on. She had been out on the concrete steps at the back of the farmhouse – slippery at the best of times – cleaning off the snow. In the cold and dark, it had made her feel like herself again.

We can all imagine a dream world in which we are rich, surrounded by luxuries and happiness. But Viktor Frankl thought we do not want happiness, we want meaning, and in this chapter, I will argue that money, like happiness, is at best a proxy for what we really want. In the last few chapters, we have looked at one of Frankl's main sources of meaning, our relationships and interactions with the people around us – family, friends and enemies. As I have argued, we gain meaning from discharging our own

IRMs – caring for our children, exchanging gifts with friends, sharing grief at a family funeral and, also, perhaps disconcerting but I think certainly true, making an enemy pay for the wrong they did us. All of these make our lives feel real and important; all make us feel right about ourselves and our place in the universe. Now, we can turn to another of Frankl's ideas, the search for meaning through struggle.

Facing a challenge may not always make us feel happy. It is hard to say that you are 'happy' when, out in the dark night, you have a freezing shovel in your hands, your knee is refusing to support you and you are shovelling snow off slippery concrete steps. People would certainly pay money *not* to do this. Then when you are back by the farmhouse fire, picking up the phone to speak to your son, you have the sense that the day was worthwhile.

The esteem of others – and especially ourselves

To understand this need for struggle, I think we should begin from pecking orders, status and Fiske's 'authority ranking' mode. In *The Descent of Man*, Darwin refers to 'the love of praise and the strong feeling of glory, and the still stronger horror of scorn and infamy'.[1] From John Adams, one of the founders of the American constitution: 'The desire for the esteem of others is as real a want of nature as hunger.'[2] We live in the regard of others, and in every corner of our lives, we strive to be judged well.

Such judgements imply status, with a scale of who has done better and who has done worse. Glory means that we are raised up above our fellows. Shame means that we have lowered ourselves below them. But we should immediately be careful with this thought. Pan doubtless gives us the hunger for status, but it is Spock, with his infinite variety of ideas, who fills in the content.

With this variety, we can value almost anything, from vegetable plots to hairstyles. As we saw in Chapter 7, we have the 'sweet potato bigman' of the Eipo, and across cultures, people who are valued as leaders in battle, builders of huts, navigators, artists, scientists. In particular, we can value many social roles. For many animals, status may largely be a matter of strength and size, but for us, at the same time as we revere the successful leader, we can praise the person who submits to the leader's authority, and diligently fulfils his or her role in the social order. While we compete for esteem and status, 'status' can take as many forms as humans can imagine things to value.

This variety is everywhere. In the playground, children have an urge to show off their skills on the climbing frame or skateboard, and back in class, they show off how well they can sing, or draw, or how quickly they can raise their hands to answer the teacher's question. Adults may encourage this pride, or they may discourage it as going too far – but both approval and pulling back indicate the intense human concern with relative standing.

Similarly, as adults, when we are proud of our sweet potatoes or our symphony, we may try our best not to make our audience feel belittled, but the element of competition is there nevertheless – without it, there would be no meaning to a fear of 'belittlement'. Just the possession of information in itself can confer status. In social gatherings, it is common to see groups of people – in my experience, usually men – chatting about a political event, or a machine, or climate change, each with a little smile on their face as they show off their (often rather obvious) knowledge. The phenomenon may not be entirely masculine, though the expression 'mansplaining' suggests that it commonly is.

Perhaps most concretely, in sports we all know the joys

of competition, struggle and, above all, victory. In my ivory tower of a workplace, where scientists supposedly assemble for arcane struggle with the puzzles of the human mind and brain, it may well be the triumph of a football team that dominates the Monday morning conversation. I have always felt utterly uninterested in athletics (what can I say, at school I usually came last), but the 1981 film *Chariots of Fire* is irresistible. In the middle of his race, the protagonist falls; as he leaps back to his feet, the look crossing his face lets you know what will happen; as he chases down the pack and crosses the finish line, first there is a swelling sense of something special, a shiver up the back and a lump in the throat.

In humans and other animals, aggression directed to other members of the species is used in competition for resources, territory and mates. As Lorenz pointed out, though, in humans there has been a quite remarkable change. In evolutionary theory, there is the concept of exaptation – a structure originally evolved for one purpose becomes adapted for another – and in human life, the aggressive drive has been exapted to make an enemy of any challenge. As we struggle for success, the aggression is turned away from human competitors to the challenge itself. We speak of 'getting our teeth into' a problem and, if we fail, the problem has 'defeated' us. Eibl-Eibesfeldt pointed out that the very word 'aggression' derives from the Latin *aggredi*, meaning not only to attack but to approach, tackle, undertake.[3]

Research shows that the facial expressions of anger and determination resemble one another – a photograph showing strong determination may also be judged to show anger.[4] In a fascinating study from the psychologist Michael Lewis and colleagues, infants aged five months learned that pulling a string produced a colourful slide and music.[5] Then, conditions changed and the pull stopped working. Some infants responded by getting angry, others by getting sad. Assessed again at age two, the infants who had got angry as babies were now more likely to persist if their play was temporarily interrupted.

We are so used to this link between challenge and anger that we do not realise how remarkable it is. If an obstacle to progress is really severe, actual fighting can break out between ourselves and an inanimate object. 'Getting our teeth into' a mathematical problem may seem like just a metaphor, but sometimes we do actually get out our weapons and batter our 'opponent'. Hilarious video clips show men losing their temper with inanimate objects and hurling them to the ground,[6] and I have a favourite with a furious human soundtrack accompanying a cat who is simply playing with a recalcitrant printer.[7] This second one beautifully shows what is special about human aggression. The cat is not funny, and the cat is not angry. There is no expanded tail, arched back, hissing and howling. A kitten may sometimes launch a play attack on a ball of string, but it is simply inconceivable that a cat would genuinely fly at a printer, or any other inanimate object. It is the human voice that changes the story to a story about frustrated rage . . . in this case, a story that, for all of us, is only too horribly familiar.

The critical thought is that, for aggression to drive us forward, it does not have to be war, it does not have to be inflicting injury. Just the same IRMs are in play as we open the farmhouse door on a black winter's morning and discover we shall be fetching the cows to be milked through a roaring hurricane, or as we struggle with the equations in a scientific paper and realise we have still not really followed the author's argument. The opponent is no longer another human being. It is the challenge to be met and subdued. For this perhaps more than many other reasons, I believe (again, along with Lorenz[8]) that it would be a really bad idea to try and eliminate aggression from the human psyche. The right move is not elimination, but channelling.

As John Adams said, we hunger for the esteem of others, but it goes deeper than that. As we grow up, the esteem of others gradually transforms into esteem of ourselves. As I struggle to write a scientific paper, of course I want it to be greeted well by the community. Indeed, if it is not greeted well by the community, there are practical consequences – my paper is not published, my grant money may dry up, my colleagues may pass me by for promotion. Deeper than this, though, is the satisfaction of feeling that the paper has been done well. I wish to be proud of it in front of my peers but, even more, I wish to be proud of it in front of myself. I have often been struck by an actual physical sensation that floods through me when I have been sitting thinking, usually about some problem in psychology, and finally come to a solution that seems exactly right. It is something bodily, flooding down through my chest and abdomen – an actual physical pleasure at having got it right.

Many years ago, I was listening to a talk on job satisfaction. I no longer remember who gave the talk or whose research was being described, but I do remember that two dimensions predicted which jobs were satisfying and which were stressful. One dimension was uncontrollable time demands – too much to do, too little time, bad. The other was the degree to which the job afforded opportunities to exercise individual skill and ability. (Again, an idea strongly reminiscent of Viktor Frankl, though presumably arrived at independently.) Using these two dimensions, jobs could be plotted in a space showing how stressful or satisfying they were. In the bottom-right quadrant were the worst jobs, with time pressure and little sense of individual skill – perhaps work on an assembly line, with car parts incessantly passing and nothing to do but attach a particular nut and bolt. In the top left was the perfect job, which was 'forester'. What I remember is sitting in the talk thinking, *Hell! Why didn't I become a forester?!*

The need to struggle and succeed captures one aspect of Viktor Frankl's thinking about the sense of meaning in life, and very similar is his take on suffering. We have a deep need to show others and ourselves that, when things get hard, we can take it. For males especially, many cultures have used painful initiation rites to mark the transition from boy to man. In Auschwitz, Frankl saw the difference between the prisoner who awoke to meet the challenge of another day in the Polish winter, working without food or boots to build a railway line across the snow, and the prisoner who had decided that at last it was enough:

> Usually this happened quite suddenly, in the form of a crisis . . . We all feared this moment – not for ourselves, which would have been pointless, but for our friends. Usually it began with the prisoner refusing one morning to get dressed and wash or to go out on the parade grounds. No entreaties, no blows, no threats had any effect . . . There he remained, lying in his own excreta, and nothing bothered him any more.[9]

The idea of self-esteem has played a major part in psychological thinking at least since William James, the great nineteenth-century thinker whose *Principles of Psychology* did much to frame the questions of the next hundred years. James defined self-esteem as a general belief in one's own ability to succeed in the face of life's challenges.

In the most popular questionnaire used to assess self-esteem, the Rosenberg self-esteem scale, people answer questions transparently referring to success and standing.[10] From strongly agree to strongly disagree, how do you rate, 'All in all, I am inclined to feel that I am a failure,' or, 'I am able to do things as well as most other people'? In the enormous research literature on self-esteem, one perhaps rather obvious result stands out. People with higher self-esteem are happier.[11] In his highly influential hierarchy of human needs, the psychologist Abraham Maslow put needs for self-esteem, confidence, achievement, respect near the top, to be pursued once the lower needs of food, sex, safety, friendship were satisfied.[12] The same theme can be seen in the personality theory of David McClelland, with three core needs being affiliation (Frankl again), achievement and power (see Chapter 15).[13]

To be valued by others and by ourselves, we need to do well. It can be sweet potatoes, it can be symphonies, it can be equations, it can be something as pointless as football or chess – we need to struggle with the challenges of life, and to drive the struggle, I think we use the same aggression that we use to compete and struggle against other human beings. It is easy to see, too, why human life has to be this way. Our lives are staggeringly more complex than the lives of other animals, who may need to compete just for territory or mates. We can take on any goal, and a humanity that did not struggle towards its goals would be a humanity with little future on the planet. It is only sensible to suppose that human evolution needed a strong appetite for success and achievement, an appetite that could be attached to any goal – and in the usual way of evolution, this appetite, I suggest, was built on something already in place, the aggressive urge to struggle and to defeat.

Across cultures

It is sometimes thought that 'self-esteem' is a peculiarly Western or even American concept, emphasising the drive to be seen as

better than the average. A contrast is drawn with the more col-
lectivist societies of East Asia, especially Japan. Is the emphasis
on self-image, struggle and victory really specific to one culture?
In an insightful paper called 'Is there a universal need for positive self-regard?', published in 1999 in the *Psychological Review*, a
group of American and Japanese psychologists argued that the
need for self-esteem is not universal at all.[14]

The evidence assembled in this paper certainly makes fascinating reading. Where American ideals favour individualism,
the Japanese concern is focused far more strongly on the collective good. Americans are more likely to value individual talent;
Japanese are more likely to favour effort (*doryoku*), perseverance
(*gambari*) and endurance (*gaman*), wishing to show they put
in everything that they could for their community. Americans
value the glory of individual success; Japanese fear the shame
of failing their group's needs. American adults are more likely
to praise their child's strengths and successes; Japanese adults
are more likely to encourage self-reflection and self-criticism
(*hansai*). On a standard self-esteem scale, Americans score higher
than Japanese, suggesting that they are generally more pleased
with themselves.

These are big differences but, to me, they do not really
question a universal need to strive for regard from both one's
fellows and oneself. Striving is evidently central to concepts
such as *doryoku*, *gambari* and *gaman*, and the judgement of
self and others is central to a culture with a horror of shame.
It seems much more sensible to say that both Americans and
Japanese are striving to be judged well, but with somewhat different values determining how that judgement will be made.
The Japanese schoolchild may be given a horror of showing off,
and appearing to outdo his or her peers – but he or she still
strives to be judged a good member of the community. In much

the same vein, in Western culture, the sense of positive self-regard can be attached just as strongly to support of another as to individual success. There is the sense of being 'a good father' or 'a good neighbour', and when we give medals for bravery, very often they are given to the soldier who risked his or her life for a comrade. As I have said, we use many different scales to judge whether we are 'doing well', but for each scale there is the implication of an associated status, with some people's achievements superior to others.

As always with the dictates of Pan, when we look at other cultures, even those quite different from ours, we understand immediately what is happening as events unfold. Here is Eibl-Eibesfeldt on the competition for status in Trobriand Islanders:

> Each family attempts to harvest as many yams as possible, particularly large ones. The piles of yams are placed on display in the village, and the village chieftain judges the result. Whoever collected the largest harvest enjoys the highest esteem. The tubers are then stored in private storage houses, which are constructed so that the tubers can be seen between the beams; the art of storage consists of laying the most beautiful ones, so they are visible . . . Families consume as little of these yams as possible so they remain visible for a long time.[15]

With us, it may be an Aston Martin displayed in the front drive, but it is hard to imagine that it is not the same IRMs underneath.

You have to earn it

Perhaps more specifically American has been the tendency to turn things around, and to see self-esteem as cause rather than effect. Beginning in the 1970s, a swelling American movement saw low self-esteem as the cause of all manner of personal and social ills – poor educational achievement, psychopathology, crime, unwanted pregnancy. The thought was that, if society could only promote self-esteem, all these ills would diminish. In 1986, the governor of California funded a Task Force to Promote Self-Esteem and Personal and Social Responsibility, aiming through research and social intervention to attack the curse of low self-esteem.

A large body of research followed this early optimism and, unfortunately, it showed that the optimism is false. A careful review of the large research effort on self-esteem and educational success, published in 2003, concluded that the two are indeed related – at least sometimes, and even then only weakly.[16] The evidence suggested, however, that the direction of cause and effect was opposite to the direction hoped for – to a very modest degree (hardly surprising, if we think of the adolescents we know!), doing well at school may boost self-esteem, but having high self-esteem does not boost school success. We judge ourselves well *because* we do well – on whatever scale of 'doing well' we and our peers value.

We feel good when we try, and we try when we feel good

One factor linked to both effort and self-esteem is mood. The link is vivid in mental health conditions. The bible for definition of these conditions is the *Diagnostic and Statistical Manual of Mental Disorders*, published by the American Psychiatric Association.[17] In its definition of mania, we see a familiar combination – 'elevated, expansive, or irritable mood', 'increased goal-directed activity', 'inflated self-esteem'. With exaggerated belief in themselves, the person has boundless energy, pursuing goals that can be harmful or unrealistic – perhaps dangerous investments or sexual escapades. In depression it is the opposite. There is the familiar picture of sad and hopeless mood, but at least as serious are 'markedly diminished interest or pleasure in all, or almost all, activities', along with 'feelings of worthlessness'. In a deep depression, the person feels they are without value – with no credit for anything good in their lives, and deep guilt for anything bad. And they are not willing to try . . . given a choice between an easy task bringing low reward, and a harder task bringing higher reward, a person in a depressive episode is more likely to take the low-effort option.[18]

Sometimes as I cycle to work in the morning, I realise I am in a bubbling good mood, ready to race to work and get stuck into whatever is waiting. I wonder why, and realise . . . the sun is shining! It is not just me. In a paper called 'Good day sunshine' (the Beatles knew this too, of course), the economists David Hirshleifer and Tyler Shumway showed that, across 26 countries, stock markets did better when the sun shone.[19] The sun shines, we are in a good mood, the future looks bright and we're ready to gamble.

It's not really the money

As we struggle and compete, what are we competing *for*? In the modern world, perhaps the most obvious answer would be 'money'. Wealth is a very large element of social status, as we congratulate ourselves with expensive dinners, luxury cars and, in the extreme, a mansion in Mayfair and the owner's box at a sports event. We spend an enormous proportion of our time and effort on success in the best-paying job we can get.

At the same time, everybody knows that a job bringing only money can be profoundly unsatisfying. Money 'isn't enough'. In *Utopia for Realists*, Rutger Bregman cites a survey of 12,000 professionals reported in the *Harvard Business* review.[20] A full 50 per cent said they felt their job had 'no meaning and significance'. A full 50 per cent said they were 'unable to relate to their company's mission'. The jobs may pay well, but they bring a sense of just turning the wheels, with no truly satisfying product. A teacher who sees the eyes of a young child light up with understanding and enthusiasm is going to feel the day was worthwhile, but many well-paid jobs offer nothing like this. Following an article by the anthropologist David Graeber, Bregman calls them 'bullshit jobs', and argues that, in our civilisation, bullshit jobs are constantly on the rise.

A great deal of research has asked whether being richer makes you happier. By and large it does, though the richer you get, the more you need to gain in order to feel a little bit better still. In 1974, the economist Richard Easterlin described what is now known as 'the Easterlin paradox' – though wealth in a country increases substantially over the decades, happiness apparently does not, at least not in countries that were relatively prosperous to begin with.[21] The common explanation is that it does not only matter how rich you are – it matters how rich you are *compared to others*.[22] As a country's overall prosperity increases, everybody has more, but people do not become much happier if their relative position stays the same.

This is an extremely telling observation. It shows that, at least beyond some minimum, our attention is not simply focused on the things that our wealth brings us – the dinners, the cars, the private island. We may think we are working 15 hours a day to acquire these things, but to a large extent we are not. As many people have pointed out, the things are not the real point. The point is how we feel about ourselves – our sense of how well we have done and our place in the order of things. Wealth is just a clumsy proxy for the success we really want.

The media magnate Sumner Redstone rose from a poor background in Boston's West End in the 1920s and 1930s ('Our apartment in Charlesbank Homes . . . had no toilet; we had to walk down the corridor to use the pull-chain commode in the water closet we shared with the neighbors') to become a media mogul, the billionaire founder and chairman of Viacom.[23] At one point, Viacom's acquisitions included Paramount, CBS, Blockbuster, Nickelodeon, Simon & Schuster and more. The clue is there in the title of Redstone's autobiography: *A Passion to Win*. Chapter 1 begins, '*Viacom is me*. I have a love affair with this business and this company.'

The paragraph goes on with his delight in the 'global competitive struggle' to create and sell successful books, films and television. At school, Redstone is driven by an unwavering desire to be best at everything. He misses a syllable in a regional spelling bee, and writing almost 70 years later, 'I still remember the devastation as if it happened this morning.' When his takeover of Blockbuster goes wrong, his company is haemorrhaging money, but this is not what hurts him as critical notices begin to appear in the business press. 'I had come from nowhere . . . and worked hard all my life to achieve my position. Now everything was wiped out. All of my accomplishments were ignored.' And most tellingly:

I was losing money, but money was not the issue. I realise that sounds strange from someone in a position of considerable wealth, but money is really only the report card on what you accomplish. I don't believe most truly successful people are driven by money; what they want is to be the best at what they do.

Or we can look at Henry Ford's *My Life and Work*.[24] This time it is not so overtly about competition – in fact, Ford is sharply critical of what he calls 'destructive' competition, where the aim is simply to do the other guy down. Again, though, it is most certainly not about money – it is about doing the best according to one's own values. For Ford, the value is work, commitment and production.

Ford, born in 1863, also began from a poor background, this time on a farm in Michigan. As a teenager, he already loved machinery – he had a passion for taking apart and repairing watches ('At one period . . . I think I must have had fully three hundred watches'), and having been inspired by an early encounter with a steam engine, began to imagine that engines could be designed 'to lift farm drudgery off flesh and blood and lay it on steel and motors'. (Thirty years later, the first Ford tractors were rushed into production as a response to England's food shortages during the First World War; 5,000 tractors were shipped and Ford says, 'The officers of the British Government have been good enough to say that without their aid England could scarcely have met its food crisis.')

Ford transformed society with his aim to produce a car that was not an expensive luxury item, but a practical product for the common person. He pioneered mechanisation, the assembly line and mass production. He believed that a business should constantly strive to produce the best possible product at the lowest possible cost; while others increased prices, his goal was always to decrease them, and while others wished to cut the wage bill, he believed that the best business paid the highest possible wages.

Ford became one of the world's wealthiest men but, over and over again, he insists that business cannot be about making money. In the 1890s, serving as chief engineer to the Detroit

WE DON'T WANT TO BE RICH AND HAPPY

Automobile Company, he 'found that the new company was not a vehicle for realizing my ideas but merely a money-making concern'. Ford resigned, and formed the Ford Motor Company.

Elsewhere in his book he says, 'Business as a mere money-making game [is] not worth giving much thought to.' Instead, Ford's values were work, commitment and constant imagination and improvement in the aim of production. He did not allow titles in his company: 'When a man is really at work, he needs no title. His work honors him.' He believed that a person who is ready to settle back on his successes should leave business immediately. 'If the whole attention is not given to the work, it cannot be well directed.' Every success, he thought, was simply an opportunity to do more. 'The highest use of capital is not to make more money, but to make money do more service for the betterment of life.'

The point is that, as we strive for success, it is our own values that determine what 'success' will count. We wish to win, sometimes against others or sometimes just by our own standards, but the race has to be one we care about. Pan can provide this sense of purpose – the child who suddenly understands the teacher, the colleague who gives us a congratulatory pat on the back – with Spock as usual riffing on Pan's themes – victory not on a real battlefield, but on a little 8 x 8 board of black and white squares. When we simplify the whole thing to money, though, we are likely to cross the finishing line, panting, victorious, wondering why we bothered.

Running out of control

The passion to win may drive humanity to its greatest achievements, but it also runs out of control. Spock has the ability to create an infinite number of new challenges, each with the potential to raise our standing in both the eyes of others and the eyes of ourselves. As the joke goes: nobody on their deathbed ever said, 'I wish I'd spent another day at the office' . . . but this is how we live our lives, pursuing new successes long after Pan has lost interest in the things these successes bring.

Many years ago, reading *Ancient Evenings* (1983), Norman Mailer's rather brilliant immersion in the mindset of the ancient Egyptian empire, it came to me how very limited is the human imagination. The pharaoh has absolute power, with access to any conceivable product of the human mind in ancient Egypt, and what does he want? Delicious food. A harem of desirable and available sexual partners. Quality accommodations with splashing fountains. Adornment with things that shine. With absolute power, the pharaoh's top choices are very much the same as the top choices of everybody else . . . and for that matter, a list not too different from the top choices of a magpie.

In my own life, by lunchtime I am always staggeringly hungry, and when I bite into my sandwich, my body is filled with happiness. If instead I go out to the fanciest restaurant I can afford, and have an exquisite dinner, I like it fine, but my sandwich lunch makes me just as happy. Compared to the sandwich, the pharaoh and the wealthy of today may search for refined, exquisite pleasures . . . but my suspicion is that they are more exquisite, but probably not more pleasurable. When I eat my lunch, or have sex yet again with my partner of 40 years, or give another successful scientific talk, my sense is that my pleasure is already pretty close to maxed out.

I would not say that there's no point in becoming a billionaire but, perhaps, almost no point, and surely as the billions pile up there can be no more real use for them, except as a symbol of the very piling up. Once the world of Spock assembles people into corporations, the same is arguably true at the level of the new monster. An employee struggles for recognition from their peers, now in an enterprise measured by success of the corporation. At the top, success is measured by takeovers and global reach into the marketplace. But for any person working for Google or Shell, up to the board and CEO, why would it matter how many billions are on the balance sheet? It can only matter, I believe, because this is Pan run out of control, imposing a need for competitive struggle that is independent of the actual resource that is won. Again, Lorenz puts it well:

> The rushed existence into which industrialized, commercialized man has precipitated himself is actually a good example of an inexpedient development caused entirely by competition between members of the same species. Human beings of today are attacked by so called managerial diseases, high blood pressure, renal atrophy, gastric ulcers and torturing neuroses; they succumb to barbarism because they have no more time for cultural interests. And all this is unnecessary, for they could easily agree to take things more quietly; theoretically they could, but in practice it is just as impossible for them as it is for the argus pheasant to grow shorter wing feathers.[25]

(Lorenz has just been discussing the sexual selection pressures that have led to increasingly long display feathers in the pheasant – feathers that serve to attract a mate, but make it harder actually to take flight.)

Why don't we just give it away?

In a Utopian world, might we struggle for the billions as a symbol of our success – then give them all away? That would be very

Spock, and for billionaires like Bill Gates, something like this can happen. Strong forces in Pan, however, tell against this apparently rational solution. In my academic world, most people wear a socialist armband, but they struggle in every way they can imagine to pay less tax. You can say this is hypocritical but, in the conflicted world of Pan, hypocrisy is inevitable.

This goes back to the issues we examined in Chapter 7. In a mode of communal sharing, it is easy to give things away. Indeed, nothing is stronger in eliciting the attempt to pay less tax than the urge to hand one's possessions to one's children rather than the state. But we are not always in the communal sharing mode, and we have IRMs too that enforce the right to possess, to keep one's possessions and to be fairly repaid for our efforts. Evolution has ensured that we are keen to support others ... as long as we trust them to support us too; and, as Jonathan Haidt argues, the person on the left may see the poor as friends to be supported, while the person on the right may see them as undeserving freeloaders, trying to get something for nothing.[26] It does not matter that a billionaire may have nothing left to spend yet another billion on. It still will not feel right to just have the whole lot confiscated and shared out to everybody.

We have to be this way

The trouble with human conflicts is that, often, both sides are part right. Before his retirement, I would often discuss life and the mind with my friend and colleague John Teasdale. John is something of a Buddhist, whose work has been enormously influential in the development of the 'mindfulness' movement in clinical psychology. He made me read the American Zen scholar Jon

Kabat-Zinn, with his cautions over ceaseless human endeavour, and his example of the man who finds himself late on a Sunday night, still struggling to clean his car because that was on the list of goals for the weekend.[27] Of course, I recognised myself, though I am actually too lazy to let this extend to Sunday night.

John would promote the strengths of Buddhism's second noble truth, that the source of suffering is desire. I would argue, as I did earlier in this chapter, that a humanity that did not wish to struggle would be a humanity that did not get far. I still stand by my side of the argument; it is all very well for a monk to beg for food, but somebody has to be the one who sweats and produces it. At the same time, it is obvious that John was right too. All too often, we are struggling for things we could sensibly do without – not because, on Monday morning, it will really matter whether the car is sparkling or not.

I had my tongue in my cheek when I called this chapter 'We don't want to be rich and happy'. Of course, we do want to be rich and happy, but surely these are not what we most deeply want. We want the fulfilment that comes from our relationships, our struggles and our successes. We want to have lived well according to our own values.

As recounted by Herodotus in *The Histories*, the fabulously wealthy King Croesus, ruler of Lydia, asks the sage Solon who is the world's happiest man. He expects it be himself, but Solon names an unknown man, Tellus, who raised virtuous sons, saw the birth of grandsons, died valiantly in battle for his native Athens, and was buried in honour and remembered with gratitude. Next, he names the brothers Cleobis and Biton of Argos, who were to drive their mother to a festival at the temple of Hera. The oxen did not come in from the fields to drag the cart, so the two young men dragged it themselves. At the festival they were acclaimed for their strength and filial piety; the delighted mother prayed to Hera for the greatest blessing her children could receive; they fell asleep after the feast and did not awake.

Solon argued that we cannot know whose life turns out best until he or she is dead but, at the same time, the appeal of the story shows that it is our values and struggles that fulfil us, not the wealth they may deliver. The joy of achievement rests on Pan, with his aggression exapted for meeting any challenge, and his strong promptings of which challenges it is worthwhile to meet.

PART IV

Sex

CHAPTER 12

You only have to look

Two kinds of equality

In the modern world, few things are changing faster than the relations between men and women. In the 1960s and 1970s, a genuinely new kind of freedom had been created by introduction of the birth control pill. It may be hard now to recapture the feeling of change and opportunity, though you could try Loretta Lynn's song 'The Pill' (few people knew better than Loretta Lynn how to capture the voice of the everyday woman), and for her delicious comment on culture clash, you could add 'One's on the Way'.[1]

With this sense of their lives in their own hands, it seemed women were ready to grab their newfound pride and power. The feminist bible was Germaine Greer's *The Female Eunuch* (1970), and if you look back at it now, it is perhaps surprising. There is a good bit about men's suppression of women, but the stronger theme is that women need to stop dreaming about the perfect man and get out there to take what is theirs.

In 1976, I was a postdoc in Eugene, Oregon, and Sharon Posner, the wife of our lab head, handed out T-shirts in assorted bright colours, saying this: 'A WOMAN'S PLACE IS IN THE HOUSE [image shows typical North American single-storey suburban dwelling] ... AND THE SENATE [image shows US Capitol]'. (For those non-Americans who may be oblivious, an association with SENATE very much changes the meaning of HOUSE.) The T-shirts were extremely popular. My wife wore hers for many years, until the purple faded pretty much to nothing.

I still love the inspiring sense of newfound opportunity from this time. I love its redirect jokes: 'Why are blonde jokes so short?' 'So men can remember them.' Or the one I quoted earlier: 'What's the name for that useless piece of flesh at the end of a penis?' 'A man.' Or for a recent version, the PopcornMax sketch of hapless males blissfully unaware of how things vanish from the laundry basket or the 'magic' coffee table.[2]

But, as feminism has become more academic and intellectual, it has become more than a demand for equality and opportunity. In his profound study of kibbutz women, *Gender and Culture*, the anthropologist Melford Spiro lists four axioms of academic 'women's studies':

(a) gender and gender differences are culturally constructed;
(b) the mother's tie to her child is cultural, not natural; (c) female liberation can be achieved just in case the family and child rearing are removed as a focus of women's interests; and (d) the abolition of sex-role differentiation is a necessary condition for sexual equality.[3]

In Chapter 14, we will come back to Spiro's work, and his reasons for doubting all these axioms. Meanwhile, such axioms sit uneasily with the complex world of Pan and Spock. Certainly, culture is a large force shaping our ideas of how men and women should behave – but there is the animal too and, in the context of biology, it is highly unlikely that we can bring up our boys and girls to be the same people. Culture evidently influences sexual attraction and sexual activity – but not only culture and, when it comes to sex, Pan tends to follow his own rules, rather independent from Spock's advice. In line with Spiro, there is strong reason to doubt that family life, with its deep absorptions and deep satisfactions, is just a political and cultural institution, to be redesigned as we wish. As we will see from his work, the family means something, and when we decide it does not, we design a life that is unwanted.

As for power, regard and reward, it is obvious that discrimination of men against women, and of women against women, can be an entrenched cultural institution – but in the world of Pan and Spock, biased beliefs and practices are only one part of the story.

Once again, of course, I do not believe that any of this is simple. It is the usual chaos of Pan and Spock, and when we insist on Spock's over-simplified ideas, as always, we miss much of the truth about ourselves, and the heady mix of many things we need and want. There is the animal as well as the idea, and if I am right that much of our sense of meaning comes from the animal, then much will be missed by an attempt to exclude it from men's and women's lives.

Needless to say, many feminist writers have questioned the 'axioms' listed by Spiro. The feminist sociologist Alice Rossi described gender study that does not deal with biology as 'an exercise in wishful thinking'.[4] In her 1994 book *Who Stole Feminism?*, the philosopher Christina Hoff Sommers distinguished equity feminism from gender feminism.[5] Equity feminism is a call for fairness, for equality of treatment and opportunity. In gender feminism, a core belief is that differences between men and women are a cultural invention, designed to perpetuate male power. It is gender feminism that Spiro is characterising, while in the 1970s, when we all loved the T-shirts, it was equity feminism we had in mind. While we hoped for a world in which men and women might both grab their chance for the House, it never occurred to us that men and women were actually the same.

Equity feminism is a demand for a new social order. It is all the best of Spock – the cultural situation changes, ideas change and there is a reweighting of social values. Referring back to Fiske's moral modes, we might see this as a change in authority ranking, a reversal of the assumption that men have precedence. Equity feminism rebels against 'the patriarchy' – that is, the whole social structure that keeps power and privilege away from women. It is the feminism of Susan B. Anthony and the women (and sometimes men) who struggled through the nineteenth and early twentieth centuries for women's right to vote.

In contrast, gender feminism involves a claim – in my opinion, a rather far-fetched claim – about facts of human nature. In this case, once again, we see the tendency for Spock's over-simplified ideas to run away with themselves – especially, perhaps, among academics. The starting point is a political aspiration, a demand for freedom from domination, but in the search for abstract 'equality', we end up with conclusions that are divorced from reality.

In the Introduction, I recommended Chimamanda Ngozi Adichie's 'many stories' lecture, and just as powerful is her 2012 lecture 'We Should All Be Feminists'.[6] There is determination, there is the power of culture, there are calls to action, but in the world of a luminous novelist, there is also complexity and the many sides of reality. In the 2022 Reith Lectures, Adichie delivered an eloquent defence of seeking the truth, even when truths have many sides and even when they are unwelcome. She put it like this: 'Literature deeply matters and I believe literature is in peril because of social censure. If nothing changes, the next generation will read us and wonder, how did they manage to stop being human? How were they so lacking in contradiction and complexity? How did they banish all their shadows?' In 2015, it was announced that the written version of *We Should All be Feminists* would be given to every high school student in Sweden. In contrast to academic feminism, you might call this real-life feminism.

Distributions

Whether it is Pan or Spock that creates differences between women and men, there is something to keep clearly in mind. Any difference we may consider, whether cultural or instinctive or physical, is not a matter of single people. It is a matter of distributions. Distributions look like the examples you can see in Figure 4. We take a large sample of men and a large sample of women, and for each person, we measure something. It might be the strength of their right arm, or their IQ, or how much they know about English grammar. Each person gets a score – we calculate the percentage of people who get each score, and draw a diagram of the result.

Depending on what you measure, the results could look like any of the examples in Figure 4. In the top set, distributions are exactly the same for men and women. Many years ago, I attended a conference on intelligence, and one of the main lectures was given by a man who had combined the data from several very

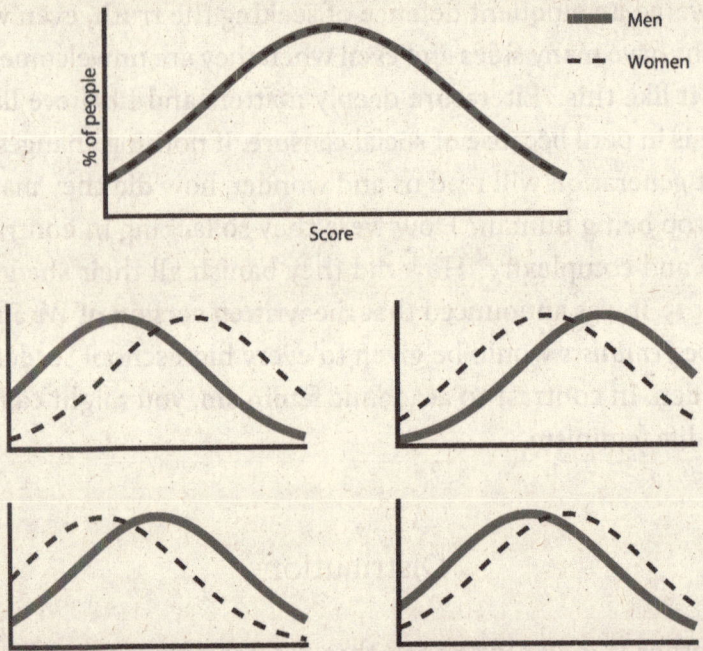

Figure 4. To produce distributions, we measure something in a large sample of men and women, and plot the percentage of people getting each score. Identical distributions for men and women are illustrated at the top. IQ looks very much like this. Four examples of different distributions are shown underneath. Any particular physical or psychological feature will have its own pair of distributions.

large samples to compare the average IQs of men and women. I don't remember the exact details, but finally he had data from tens of thousands of people of each sex and, with these enormous numbers, he was able to conclude that, according to this IQ test, men had an advantage over women – by some tiny, completely trivial amount. He was speaking about the tiny difference, but for all practical purposes, the distributions were like the top pair in Figure 4. Waves of hatred were coming back from the audience, and as a matter of fact, the speaker did move on to draw some extremely suspect conclusions, regarding things like the proportions of men and women getting first-class degrees from

Cambridge, but sitting in the audience, my own jaw was hitting the floor for the opposite reason. I was thinking, *Really? – you measure something in TENS OF THOUSANDS of men and women and still the average, as close as makes no difference, is EXACTLY THE SAME?*

When I heard this lecture, I thought – and still think – that this is an astounding biological regularity. Men and women have different genes, different upbringing, different attitudes, different experiences – and still, when you measure tens of thousands of each sex, their IQ distributions are almost perfectly matched? This is not the place for a discussion of how this can be, but it is astonishing . . . and honestly, I imagine there are very few other things you could measure that would give a result like this.

Instead, it is usually going to be something like the other four diagrams in the figure. The distributions always overlap, so you can always find a man and a woman with the same score (except perhaps at the real extremes – possibly there really is no woman anywhere who is as strong as the strongest male weightlifter). But though they overlap, the distributions also differ – 'on average' (as we say, though the average is not a great summary of the whole distribution), men score higher on some things, and women score higher on others.

Which do we care about – the individual, or the distribution? When we meet a person, we care about them as an individual, and for all the many things that matter in our dealings with that person, it is their *own*, individual characteristics that matter. Once I took a large chest freezer to the dump and asked the woman who was clearly in charge if somebody could help me lift it out of the car; she walked over to the open boot, picked the freezer out and dumped it into a skip. In this case, distributions of strength in men and women do not matter – she was herself, perfectly able to pick up a freezer that would have put my back out.

But when we talk about society at large, it is the distributions that matter. If we want to know why, for example, the enormous majority of murders are committed by men, this is a question about distributions, and the way those distributions are shaped by Pan and Spock.

Equity feminism demands that boys and girls be given the same chance to develop their talents and aspirations. It demands that, when a man and a woman grow up to be similar, they are given the same chance of success. It recognises that distributions of mental and physical features overlap so that, very often, a man and a woman will indeed have similar talents, aspirations and priorities. Gender feminism, in contrast, believes that, absent cultural bias, the distributions of men and women – at least, distributions of mental rather than physical features – would be the same. On this view, differences between men and women are all Spock, and with a revolution of our culture, all could be changed.

The usual vertebrate picture

Chimamanda Ngozi Adichie says, 'We are not apes,' and she is certainly right.[7] Every species is different, and humans are certainly unique, with all the complexity of instinct and culture in cooperation and collision. But still . . .

When I grew up, there were around a hundred cows on the farm, but no bull. There was no bull because one of the last bulls that the family had owned had come close to killing my father. On a winter's morning, as he took it a bucket of food, the bull had knocked him over on the icy ground; for several minutes he was on his back in the snow, the bull kneeling on his chest

268

trying to turn its head to gore him while he held frantically on to the ring in its nose. He always maintained that if it had not been snowy, he would have been killed, but as the weight of the massive bull threatened to crush him, he kept slithering to one side, still hanging on to the animal's nose. The bellowing of the bull and yelling of my father at last brought my grandfather and a neighbour running; armed with heavy farm brushes they drove the animal off and, soon after, it departed for good, while my father recovered from his broken ribs. But as children, we could be sent alone to bring in a hundred cows for milking.

We could also be sent to get eggs from the henhouse, and when I was very small, I was a bit discomfited by the way that a dozen hens fluttered and pecked. But one summer, our neighbour had a bantam cock that was disposed to fly at any visitor as they approached the farmyard. When we went visiting that summer, we took a stick.

We are not apes, and we are certainly not herbivores or jun-glefowl, but we are animals. Each species differentiates males and females in its own way. In the spotted sandpiper, for example, it is the females who are larger than the males; the migrating females arrive first at the breeding grounds, fight for territory and for males, and on her territory the female may keep several males, each incubating his own clutch of eggs.[8] Something similar is true of the red phalarope, an Arctic wader, where again the female is larger and more colourful, fights for males and provides each with a clutch of eggs.

Among vertebrates from fish to birds to mammals, however, these cases are unusual. By far the most common pattern is that a large, muscular, aggressive male fights for territory and displays to attract a mate; a smaller, more selective female watches and chooses.[9] As regards evolution, we understand a lot about *why* this happens (see the next section of this chapter), and as regards brain development, we understand a lot about *how* it happens (section after that).

As for *Homo sapiens*, you don't need data on murder rates to know it happens with us too. (According to the 2018 edition of the *UN Global Study on Homicide*, about 90 per cent of murders are committed by men.) You don't need to look at who is showing off and being disruptive in a primary school class. You don't need to hear parents telling children to 'look out for strange men'. You don't need research evidence showing that, across 101 cultures, boys aged two to six are physically and verbally more aggressive than girls;[10] that respondents in every one of 25 countries judged the adjectives 'adventurous, dominant, forceful' as more male, but 'submissive, sentimental' more female;[11] that as toddlers, boys leave their mothers more easily than girls, whether this is a study of Londoners in a park or !Kung Bushmen;[12] that all of this matches the typical behaviour patterns of Old World monkeys, with males showing more juvenile fighting games, displays and exploration, and females showing more social contact and care.[13] Honestly, to assess whether we follow this typical vertebrate pattern, all you really need to do is look at the size of our shoulders.[14]

Some time ago, my wife and I watched every episode of the massively charming *Inspector Morse* prequel, *Endeavour*. Some way in, we realised we could usually guess the murderer by the simple rule of choosing the most unlikely-seeming woman in the story. Of course, I do not know this, but I have to suspect that the twenty-first-century scriptwriters had decided to say 'women have the right to be murderers too'. If I were a woman, this is a right I would not be especially keen to defend, but whatever we think of the politics, this view of 'equality' certainly does little for the show's realism. In real life, it is almost never the woman.

Why it has evolved this way

Why is it that, in most cases, it is the males who are large, aggressive and prone to showing off? In evolutionary theory, the explanation is now widely agreed. Both males and females need to behave in a way that optimally sends their genes into the next generation, but for the two sexes, this calls for different strategies.[15]

For females, usually, there is a large investment in individual offspring. From insects to vertebrates, an egg is large and takes a lot to produce. A sperm, in contrast, costs next to nothing. In many animals, the female follows up the major investment of an egg with the further investment of significant parental care. This may be a sensible evolutionary strategy because it is important for the valuable original egg to be transformed into an adult. Where females mate with multiple males, furthermore, the female who invests parental care is certainly investing in her own offspring, which is not true for the male. For the female, reproductive success depends on the survival probability of her limited number of potential offspring. It makes sense to be careful in choosing the mate that will maximise this probability.

For the male, it is different. The limiting factor is not the number of sperm that can be produced; it is the number of mates that can be attracted. The ratio of sexually active males to sexually receptive females is called the 'operational sex ratio', and because males are generally available to produce more sperm, while females have less to gain by simply mating again, this ratio is usually greater than one. When this is true, males usually vary in their mating success – all are ready to mate, but only some succeed (often multiple times) – while females vary much less, since one way or another, they are all able to find a partner. This difference in the probability of success is called Bateman's principle, after classic work on the fruit fly *Drosophila* carried out in the 1940s by the geneticist Angus Bateman.

The male needs to prove he is a worthwhile partner; the female needs to make the best choice she can. What is a good choice? In part, this is a matter of resources.[16] The best male can provide the things that are needed to raise the immediate offspring – food, shelter, protection from predators. But the male also needs to provide genes that will prove successful in the next generation – for example, genes that will make the next generation's females wish to mate with the current female's sons.[17] A male that looks attractive is a good investment simply because his offspring are likely to look attractive too.

Across the animal kingdom, this sets up a scenario where males compete to win breeding territories and to attract females; where they invest in spectacular displays of their strength and vigour; where they develop formidable armaments, from the antlers of duelling beetles to the not dissimilar antlers of duelling deer. They need the armaments, and they need the aggressive temperament to use them.

Sometimes, the most striking feature is the male's display. We all know the spring explosion of birdsong and, sometimes, quite specific features of the song make it attractive. The female canary, for example, is attracted by the male's ability to pack many two-note syllables into a brief burst of song, preferably over a wide range of frequencies.[18] When the song is played artificially, especially if these features are exaggerated, the female adopts a characteristic pre-copulatory pose, indicating readiness to mate.

Of course, it does not have to be song. In Australian bowerbirds, males build elaborate structures of sticks decorated with flowers, pebbles and other bright objects.[19] The female passes from one bower to another, inspecting their quality, and then a few days later, once she has built her nest, she returns to inspect several bowers again, usually with the male in a full courtship routine of dancing and trilling. Finally, her mind is made up; she enters her chosen bower, is courted again and at last mates.

Sometimes the dominant element is not display to the female, but battle with rival males. In species where females gather in groups, from lions to cattle, one male may defend a large group or, sometimes, the attractive territory that itself draws the group. Sometimes, everything depends on the season. Males of the Yarrow's spiny lizard live peacefully together as long as it is not breeding season, but as the females become receptive, they turn to aggressive territory defence.[20]

It is usually this way around but, as I mentioned earlier, not always, and the exceptions are informative. If the male offers something especially valuable, and the operational sex ratio is reversed, then now it can be females who are large, colourful and competitive. The mating swarms of empid flies, for example, contain many more females than males, because the males are off hunting, aiming to bring back prey that will be offered as a nuptial gift.[21] When a male arrives, bringing back this valuable gift, he is surrounded by females, displaying decorated wings, legs or inflatable sacs on the abdomen. Something similar can happen with male crickets, which contribute to offspring development with a large edible spermatophore that is given along with the sperm. The spermatophore can be 25 per cent of the insect's total body weight, meaning that now it is the males that will be unable to mate with multiple partners. When the male sings to show he is ready to mate, the females jostle for his attention, and it is the male who rejects a female who is too small.[22]

There is another case worth thinking about. Sometimes both sexes are making substantial investments, and now both evolve ornamentation, sexual display and choice.[23] *Homo sapiens* may be biased towards the standard pattern of male display and female choice but, fairly obviously, this bias is far from absolute.[24] We have offspring that need many years of investment from both parents, and to a degree we have competition in both sexes, courting and display from both sexes, and choice in both sexes. It may be most usual for males to take the lead in courtship, but surrounding this are many cultural variations as well as cultural conflicts. This looks again like competing IRMs, some for display and pursuit, but some also for assessment and choice – perhaps varying in strength between men and women, but to a degree present in both. Look back at the distributions.

Hormones, body and brain

Now from evolution to biochemistry. Evolutionary theory tells us why males and females differ; biochemistry tells us how these differences come about. As everybody knows, males and females differ in their hormones – chemicals that are released from various organs in the body, circulate in the blood, bind to cells and change the way they work. In particular, hormones bind to cells in the brain and produce their characteristic effects on neural firing and behaviour.

One major player, of course, is testosterone. Manufactured in the testes, testosterone circulates in the body and, in many species, promotes masculine features of body and behaviour. Male lizards, sparrows and red deer may all live peaceably together through much of the year, and then as the mating season approaches, testosterone levels rise, the testes may grow, antlers or other weapons develop, territories are established, battles break out.[25] A tame male squirrel suddenly becomes aggressive and attacks its keeper. A castrated animal does not show this sudden change but, given hormone replacement therapy, he gets it back.

This is all strongest in males, but testosterone does not come only from the testes and, in some animals, it is related to aggression in females. In the dunnock, a small songbird, several females may live in the vicinity of a single male, and they battle with one another for the male's attention. In the females of this species, testosterone levels are higher when there are competing females to be dealt with.[26]

In some cases, we now know a lot about the complex biochemical events that lead from testosterone to behaviour.[27] After mating, the male Japanese quail appears fascinated by his new partner, spending much time staring at her through the bars of his cage. Testosterone is critical; the castrated male is not interested, but with an implant of testosterone, his staring returns. Careful biochemical analysis shows how this works. In the preoptic nucleus of the hypothalamus – one of those ancient brain regions that we considered in Chapter 4 – a particular enzyme converts testosterone into a different hormone, oestradiol. Oestradiol is often considered a 'female' hormone, though it is also present in males and, in this case, is a critical step in the chain of events. Oestradiol binds to a receptor protein on the surface of neurons in the preoptic nucleus, and after a series of biochemical reactions within the cell, the neuron fires. Firing is transmitted to output regions of the brain, very much as we discussed in Chapter 4, and now the bird approaches the side of his cage and stares lovingly at his partner.

In many animals, testosterone first exerts its effects in the developing foetus. With high testosterone, the body becomes male; with a lower level, it becomes female. Accompanying these differences in the body are differences in the behaviour that emerge later, as the animal grows up. Evidently, hormones in the developing foetus have set the stage for adult brain activity.

In young rhesus monkeys, for example, males and females play differently, in very much the way you might expect. Males are more likely to initiate play, and their games involve more pursuit, threat and rough-and-tumble.[28] But if the mother was treated with testosterone while her babies were still unborn, her female children would be more male – not perhaps as rough as the real males, but rougher than the usual female. Experiments like this show much the same in other mammals – mice, rats, guinea pigs. Testosterone added *in utero* makes female behaviour more male. Meanwhile, as first shown in chicks by the nineteenth-century physiologist Arnold Adolph Berthold, castration of young animals makes adult males more peaceable.

Testosterone also plays a major part in the differences between males and females of our species. For males, there is a burst of testosterone release in the womb, with another shortly after birth – then, until puberty, testosterone levels are about the same for boys and girls. At puberty, levels increase in both sexes, but the increase is massively more for men, whose testosterone concentration as adults will be about ten times that of women.

We cannot directly manipulate prenatal testosterone in human infants, but there are other ways to see its effects.[29] In girls, a genetic condition known as congenital adrenal hyperplasia (CAH) increases the level of testosterone in the developing foetus. As these girls grow up, they show increased preference for male-typical toys, male playmates and male activities. As adults, they are much more likely than other women to prefer a female sexual partner, and to feel that they are men. Something similar happens if, for some clinical reason, the mother was prescribed a testosterone-like hormone during pregnancy.

For boys in the womb, a complementary clinical condition is

complete androgen insensitivity syndrome (CAIS).[30] The testes of these boys release testosterone, but the cells of their body are unable to respond to it. In addition to genitalia that appear female, these genetic boys show more female-typical play and, as adults, almost always prefer men as sexual partners. Then we see similar things again at puberty. If a clinical condition reduces testosterone output in the boy, the usual adolescent changes in voice, hair, musculature and behaviour are diminished, but return with replacement therapy.[31]

Humans also resemble other animals in the way that testosterone and aggression support one another. In macaque monkeys, a strange male introduced into a group may be attacked and quite likely wounded. In this beaten stranger, the testosterone level collapses.[32] The opposite happens when a mouse wins a fight; testosterone increases.[33] In us, testosterone goes up when we watch our favourite sports team win, and goes down when we watch them lose.[34] The competitive, aggressive, 'high-testosterone' man is part of popular culture; though, in fact, a synthesis of much research suggests that any link between a man's baseline level of testosterone and his level of aggression is extremely weak.[35] (Evidently, there is a lot to aggression besides testosterone.) Aggression may be more related to the change of testosterone that accompanies some kind of competitive challenge.[36] With humans, where direct experiments are impossible, it can be hard to get at cause and effect. For example, among prison inmates, there are studies that show higher testosterone in people convicted of violent crimes;[37] but quite possibly, men convicted of violent crimes are also more violent in prison, and it may be this day-to-day aggression that itself has released more testosterone.

Of course, testosterone is just one part of the complex story of sex and hormones: oestradiol, progesterone and more, all differing between males and females, and all with effects on body, brain and behaviour. Any one of these could be material for chapter of its own, but here we can just look at one more pair of hormones, oxytocin and vasopressin.

Oxytocin and vasopressin are two very similar mammalian hormones. Both are synthesised in the hypothalamus. They are then released into the bloodstream via the pituitary gland, and at the same time transmitted by hypothalamic neurons to other structures in the brain. Across the animal kingdom, from insects to snails to worms, there are hormones very similar to oxytocin and vasopressin, with major effects on sexual and social behaviour.[38] Even in snails, for example, the related hormone affects ejaculation in males and egg laying in females. In humans and other mammals, oxytocin increases during lactation, birth and mating. In both men and women, oxytocin in linked to orgasm.[39] Across species, these hormones affect sexual and parental activity, bonding to partners and to offspring, aggression and territory defence, often with striking differences between males and females.

The details, of course, depend on the species. A great deal of work has focused on the prairie vole, an animal that forms strong, individualised bonds between male and female. Male prairie voles, unlike the males of closely related species such as the meadow vole, show strong preference for the female they have mated with, and share in care of the offspring. The key hormone is vasopressin; by genetic engineering, the vasopressin receptors in the brain of a male meadow vole can be made to resemble those of the prairie vole, and when this is done, the meadow vole now becomes a faithful partner too. For females, oxytocin is important. Blocking oxytocin receptors in the brain inhibits preference for the partner, while administering oxytocin strengthens it.

Across many species, both males and females synthesise

oxytocin and vasopressin, but vasopressin appears more important in the male behaviours of aggression, territory defence,
erection and ejaculation, while oxytocin is more important in
the female behaviours of intercourse, birth and maternal attach-
ment. In Chapter 6, we saw how, just after birth, a ewe learns the
smell of her own lamb, and thereafter will allow only this lamb
to suckle. In sheep, as in humans, there is strong oxytocin release
as a mother gives birth and, in the ewe, this oxytocin affects the
olfactory regions of the brain, promoting the selective bond to
the smell of the lamb she has delivered.

Sometimes we seem remarkably similar to prairie voles. In one
large Swedish study, with 552 pairs of human twins, variability in
a gene coding for a vasopressin receptor was linked to marital
behaviour.[40] In males only, one variant of the gene was associated with lower bonding to the partner. Men with two copies
of this variant were twice as likely to have experienced marital
problems or risk of divorce. A large number of other studies have
given people puffs of oxytocin straight into the nose, with effects
including increased gaze at the eyes of human faces, increased
generosity and trust, and reduced anxiety. Effects may differ
between women and men. Intranasal oxytocin, for example, has
been reported to increase the recognition of kinship in women
but competition in men. In part, these differences may reflect
interactions with other circulating hormones. For example,
oestradiol is known to increase the binding of oxytocin in the
amygdala, while progesterone has the opposite effect.[41]

It is not always easy to interpret the results of these experiments, since oxytocin puffed into the nose will enter the
bloodstream and circulate through the body, affecting other
organs as well as (potentially) the brain. When oxytocin
decreases anxiety, for example, is this a direct effect on the brain,
or a calming effect on the body that the brain then interprets?

Reviewing these studies, Patricia Churchland and Piotr Winkielman remind us of Orgel's third law: biology is more complicated than you imagine, even when you take Orgel's third law into account.[42] Among all this complexity, however, one clear factor is the difference between male and female. Very often, in the brain and other organs, the bodies of males and females react differently to the hormonal signals they receive.

One day, perhaps, we will be able to trace the exact biochemical pathways that, in the human brain, link hormones to behaviour. Meanwhile, we can already be sure that they are there to be traced. Hormones do not just affect the way the body forms; through their effect on the brain, they set the stage for the IRMs of courtship, aggression, bonding, childcare and certainly much more.

Shaping our children

So, does this mean that we are trapped by the evolutionary pressures that have shaped us and the hormones that circulate in our bodies? Obviously, it does not. In the world of Spock and Pan, evolution and hormones are strong influences, but they do not rule alone. In our culture, our civilisation and the way we bring up our children, we also have the rapidly evolving ideas of Spock.

Our instructions to our children may not entirely rule the way they grow up, but they do have enormous effects. We say 'Though shalt not kill', and by and large our children do not kill, despite the strong aggressive impulses released by a threatening opponent. As regards the relations between men and women, we all know the huge variations that occur across cultures and across time. In some cultures, the woman must obey the man without question. In some, they must hide their faces. In some, they are rigorously excluded from political power. In Western culture, a brief two centuries have seen staggering shifts in the roles open to men and to women.

Chimamanda Ngozi Adichie devotes a good bit of *We Should All Be Feminists* to the plea that we should change the way we raise our boys and girls, and her examples show rather vividly how expectations and choices are shaped by culture. She says, 'I know a Nigerian woman who decided to sell her house because she didn't want to intimidate a man who might want to marry her' . . . while in England rather than Nigeria, I have never met such a woman and can't easily imagine that I would.[43] Evidently this does not prove that English society is fair in its attitudes to women and men, but it does show, rather strikingly, that the way we raise our children can affect the way we think.

The way we raise our children influences their beliefs about themselves, and their hopes for what they may achieve. If we eliminate gender-biased toys and games, if we change our own beliefs and expectations, could we raise our boys and girls to be just the same? In her 1935 book *Sex and Temperament in Three Primitive Societies*, the anthropologist Margaret Mead described her studies of three groups in Papua New Guinea. The cultures she described were quite different from Western culture. She concluded, for example, that women were dominant in one group, the Chambri, while in another, the Arapesh, men and women took rather equal part in child rearing. Modern anthropology, however, has cast major doubt on these conclusions, and we now know that, across cultures, it is a very broad rule that men hold more than their share of political and religious power.[44] As noted by Eibl-Eibesfeldt, Mead herself amended her view in a later book, *Male and Female* (1949). Now she spoke of 'core gender identity' and of the psychological risks in ignoring these identities.[45]

Across cultures, besides the difference in power, there are

obvious constancies in the roles of men and women. Hunting of big game is generally assigned to men, though, as an exception, Eibl-Eibesfeldt notes the Agta of the Philippines.[46] He notes too that, in small tribal societies, children are commonly nursed for two to three years, after which another child may often be born. It accordingly makes no sense for women to join hunting bands and, much more commonly, women's tasks are focused around the village. He notes rather tartly that 'there is no tribal culture in which women are warfaring. Only in the current age of technological civilization did mankind "progress" to this stage'.[47]

Certainly, we can change things a lot in the way we rear our children, and in the way we assign power to men and women. At the same time, with the inflexibility of Pan, and the long evolutionary history of differing roles for men and women, it seems unlikely that we could turn boys and girls into the same people. In Chapter 14, I shall discuss direct evidence that we cannot.

I doubt we could turn boys and girls into the same people and, as a parent, I doubt we should try. My reasons go back to the question of individuals and distributions. When my children were quite small, we bought a cheap little picture book published by Sainsbury's, our local supermarket. The book is called *Jane and the Dragon* by Martin Bayton (1988), and as I flipped open this rather unpromising material and began to read it to my son, I suddenly and to my great surprise found I had a lump in my throat and had to pause to gather myself. Jane is a young girl growing up in a medieval castle. Everybody wants her to learn how to sew, but what Jane really loves is to practise with armour and swords. In the face of steady adult opposition, she trains herself as a knight, and one day, when all the actual knights are away at a jousting tournament, the young prince is stolen by a fire-breathing dragon. There is nobody but Jane; she puts on her armour, saddles a pony and departs to do battle. In the dragon's cave, the drama unfolds. As it turns out, the dragon never wanted to be a fire-breathing monster. It was just expected of him. There is no need to fight. They make friends. The story ends with Jane back in the castle, now instated as a proper knight, and the dragon living happily in his cave, awaiting weekend visits from his new friend.

I am still moved by this modest little child's picture book – but note what does not happen at the end. Jane lives as a knight, the dragon lives in peace, but the villagers do not decide that, from now on, all the little girls must spend as much time as the boys practising in their armour, and all the little boys must spend as much time sewing as the girls. If the book ended that way, then for me it would turn an inspiring story of individual freedom into a dystopian vision, with Spock's over-simple urge for equality crushing the children into his idea of their correct shape.

Or as another example – a while ago my wife told me about a survey she had just heard described on the BBC. Apparently, 50 per cent of girls said they didn't want to be leaders. Reasonably enough, my wife concluded (and I think the presenters of the programme also concluded) that girls should be given more aspiration. But also reasonably, I believe, I concluded that maybe we should listen to these girls. Human beings are individuals and, in my view, the parent does best by listening to their child, not just seeing them as a representative of some desired distribution.

I had two sons, and from the earliest age, they seemed like entirely different people. By the time he was five, my older son took a delight in make-up and dressing in women's clothes. He was a rather arch little boy, never afraid to engage with the adults on his own terms. In the public library in Bethesda, Maryland, one day, we were checking out books with a very large, very glamorously made-up Black woman at the desk. Peering up at her, with his eyes just about clearing the edge of the table, my son looked assessingly and said, 'Hmm . . . *orange* nail polish . . . ' – a moment's beat, and the woman smiled, stretched out her hand with fingers splayed and said, 'Yeah, honey . . . you like it?' He agreed that he did, and for a while, we would appal shop assistants, as I accompanied my son while he browsed the make-up counter and explained he was choosing which one he wanted to wear. A friend once told us that we should be careful letting him do this sort of thing or 'he might turn out to be gay' (it was the early 1990s, people still thought and even said such things) . . . to be honest, we didn't think much about it, we just thought he was cute and funny. When he was 13, he called me into his room one night and told me he actually was gay. Like any parent, I can't remotely say I'm happy with everything I did as a father, but I'm very happy that, having grown up more or less as himself, he felt at 13 that he could tell this to his father. It still breaks my heart to hear of men in their 20s and 30s who are 'not out to their parents'.

At the same age, my other son just wanted to be with the men. I remember one day when we were having our house knocked about, with bricks and dust everywhere, it was quite hard to persuade him to leave this fascinating spectacle and go to school. I told the man with the hammer that my son was refusing to leave 'the glamorous world of the builder' . . . the man looked as though he suspected me of sarcasm, but I was completely serious. In *Baby and Child Care*, Benjamin Spock (this is the other Spock) advises on how a father can help his son 'grow up comfortable about being a man'.[48] He says the father shouldn't 'jump on him when he cries, scorn him when he's playing games with girls, or force him to practice athletics. He should enjoy him when he's around, give him the feeling he's a chip off the old block, share a secret with him, take him alone on excursions sometimes'. His advice may not be perfect for everybody, but it was perfect for my second son, and I am hardly the only one to notice that, very often, children want to spend time with the same-sex parent. (Benjamin Spock goes on to talk about the father's role with girls: 'In order not to feel inferior to boys, she should believe that her father would welcome her in backyard sports, on fishing and camping trips, in attendance at ball games, whether or not she wants to accept.' This Spock, like Chimamanda Ngozi Adichie, believes in real-life feminism, with the girl free to choose what she likes.)

We can affect our child's interests, values, aspirations – but the child also brings his or her own agenda to the table, and in my opinion the good parent tries to find out what it is. Many girls and many boys will bring similar agendas – but surely both biology and our own experience tell us that, very often, they will not.

If I am asked what I think is the best ever TV programme, I usually choose *Seven Up!* and the eight (to date) sequels that have followed every seven years since. In the first programme, aired in 1964, seven-year-olds were drawn from every part of British society. The aim was frankly political, to let these children talk about themselves, their lives, their aspirations and to show the jaw-dropping differences that Britain offered to the privileged and the poor. It is so very jaw-dropping not just because these differences are so immense, but because seven-year-olds tell the truth, from the wealthy boy telling us which Oxford college he plans to attend, to the orphanage boy looking lost into the camera and asking, 'What's a university?'

Though the show was originally planned as a one-off, its brilliance proved unstoppable, and every seven years there has been a new version, following these same people as their ages progressed to 14, 21 and so far to 63. Michael Apted, an assistant in 1964 and the director of the remainder, once made an interesting comment about the show's evolution. He said that, at some point, the *Up*s stopped being political and started to be personal. Make no mistake, the political message was right. By and large, the lives of these very different children, with their staggeringly different starts in life, do turn out in the way you might predict. But they also turn out to be *their own* lives, and if I take one message from the series, it is that being rich and liking your life are two entirely independent things. We do shape our children, we do give them opportunities, values and aspirations, but they are also themselves; and surely our best bet is to understand the kind of life that this individual person would like and, where we can, give them the confidence and the opportunity to grow into it.

Where does this leave us with the thought that we should eliminate gender-based toys and frilly dresses, or avoid painting the baby's bedroom pink or blue? Are we leading the child's personality, or following it? To the extent that we are leading, we may well wish to avoid shaping boys to be boys and girls to be girls. To the extent that we are following, the boys and girls may both be happy to see their differences marked. Of course, we are both leading and following, making the choice not so easy.

In the world of Pan and Spock there are rarely easy choices but, to a very large extent, I think, it is Pan who gives us our deepest selves and our dreams. If this is true, both parent and child need to listen.

Against 'isms'

I was once sitting in the garden at work, having coffee and quite illicitly listening to the conversation between two women sitting next to me. One was describing a female friend of hers who was a serious mountain climber. She said that her friend much preferred to climb with women because they took less crazy risks. The other woman (a colleague I am rather fond of) said this was 'sexist'. I didn't join in, because I wasn't supposed to be listening anyway, but I thought this was rather unfair. I think I know what my colleague meant . . . people are individuals, and surely there are women who will take just as crazy risks as men. But still, when we think of distributions, it is a definite fact that men are more likely to take crazy risks. (A glance at statistics on traffic fatalities is enough to make the point.) And if a woman who has the guts to climb mountains wants to keep herself alive by thinking about the distributions, then I wouldn't be the one to criticise her for it.

As we saw in the case of organised religion, Spock's simple ideas often make him judgemental. He feels the rules must be followed, and forgets that his rules capture only a part of complex reality. *The Female Eunuch* has much inspiring material on self-confidence, freedom and the demand for equality, but it also has a surprising amount of criticism for women who read romance novels, or dream about a family with the perfect husband. It may be a bit cruel to say so, but my feeling is that Germaine Greer was pretty much opposed to women who did not think like Germaine Greer.

In general, I am rather allergic to 'isms'. Isms are pure Spock – under one simplifying word are swept a multitude of different beliefs and attitudes, and for any ism you can mention, I usually find that I am in strong agreement with some of those beliefs and attitudes, and in strong disagreement with others. I do not believe in socialism, or capitalism, or feminism, or sexism. Every time we use an ism, I think it would be better to take a deep breath and think what it is we're really talking about. In the case of sexism, I am firmly in favour of abolishing discrimination wherever we find it, and in giving both men and women the same opportunities to realise their dreams. I do not think this means that men and women are the same people.

Instead, I think our minds contain many IRMs, each with its

many cultural elaborations. They cover the many sides of ourselves that have appeared in previous chapters – the need to
care for a baby with a high forehead and large eyes, the need to
be treated fairly, the need to struggle against a challenge and
the shiver that accompanies it, the need to bow to an accepted
authority, and many more. As Lorenz proposed in his parliament
of instincts, these needs compete to control the things that we
feel and the things that we do, with the strength of each competitor varying over time, person and circumstance. The strength of
an IRM, and the meaning we feel when it discharges, may often be
matched in individual men and women, but this does not mean
that distributions in the two sexes are the same. Accompanying
this complexity we have the ideas of Spock, always too simple for
real life, often in conflict with both Pan and with one another.

As I said in the Introduction, as we struggle to invent culture
for the twenty-first century, we should remember that the struggling forces are the many sides of ourselves. For both women and
men, it is all the complexity of Pan and Spock. In the next two
chapters, I suggest how all this complexity is needed for lives that
are fair, free and, at the same time, matched to our own, individual passions and dreams.

CHAPTER 13

The jungle

Letting Pan happen

Sexual desires, sexual choices, sexual relationships . . . few things in our complex lives are more conflicted, or more fulfilling, than these. The 1960s and 1970s, when I began to be aware of sex, were an era of embracing the animal. We had *The Joy of Sex* (1972), the manual that aimed to transform sex from the dirty word of yesterday to the everyday enjoyment of relationships today. Emphasising the everyday normality, the title was a play on middle America's most indispensable family cookbook, *Joy of Cooking* (first published in 1936), and in fact, my father once told me, 'Sex is like making a good stew – find out what she likes and add more of it.'

In the many line drawings of *The Joy of Sex*, the women as well as the men have not shaved their legs or their armpits, let alone their pubic hair – the scent of a (clean) body is put forward as a powerful sexual asset, with deodorants strictly prohibited – the animal is at the heart of everything. Women discarded their bras and were for this short time proud to display the results. An iconic image is the album cover of Carly Simon's *No Secrets*

(1972) – she stands in her hippy clothes staring out at the purchaser with a look that says, 'This is who I am – who are you?' In the song that made her a star, 'You're So Vain', she is ready to take shit from nobody, pitilessly laying out the superficial self-absorption of the older man who used and discarded her. In the Notes is a link to a performance that captures the casual optimism of the times.[1]

As a teenager, I was horribly (and unrequitedly) in love with a girl who used to enjoy terrorising me with her sexuality. Looking at Carly's braless image, she told me she wished she had what Carly had. At about the same time, Carole King sold 25 million copies of *Tapestry*, singing the song she co-wrote '(You Make Me Feel Like) A Natural Woman' (in the Notes is a live performance of this too).[2]

For a few years, 'sex' was about the most unambiguously positive word of the English language, and it was all about the animal.

Needless to say, it didn't last. Bras came back, razors came back, deodorant came back. A few years ago, my wife and I were stunned to hear a radio programme in which young women all admitted to shaving their pubic hair – they said they were afraid that otherwise their boyfriends would think them dirty. In 1970, we thought the (to us) prudish rejection of the animal was gone from our culture. We were wrong.

In the 2020s, we speak more easily of 'gender' than of 'sex', and the choice of word is interesting. It was sociologists and psychologists in the 1940s and 1950s that first used 'gender' in reference to human beings. Before this, 'gender' was a term from linguistics, referring strictly to *words* – to the male and female of French *le couteau* and *la fourchette* – not to animals (people) but to symbols. Now, perhaps we prefer to emphasise the idea over the animal – to people as they see themselves, not as biology dictates – but I sometimes wonder whether the word 'sex' reverted in our minds to its slightly dangerous previous state, promoting a search for something safer. (I even sometimes wonder if this happened because of AIDS – certainly a mood of eager optimism must be at least discouraged by the need to take precautions against a potentially fatal infection.) Perhaps it is reassuring to think of ourselves not as messy animals but as clean, organised social symbols. But whatever we are, we are not symbols.

What we are is the usual patchwork of conflicting innate releasing mechanisms, thoughts and rules. Like the dancing male mantis or the bowing pheasant, we have elaborate courtship rituals – the flirting, the eyes looking up and down, the gifts – in our case expanded with the many details of different cultures. A psychologist friend of mine calls this 'the dance', and in courtship, the dance is everything. (The feminist scholar Andrea Dworkin once said 'seduction is often difficult to distinguish from rape. In seduction, the rapist often bothers to buy a bottle of wine'³ . . . but the dance is surely as important to courting humans as it is to the mantis, who gets it wrong at peril of turning too soon into food.)

We are pulled in different directions by all the IRMs you would expect of a species with a muscular, displaying male, a selective female, a need for long-term childcare, an instinct to preserve the sanctity of the culture, and more, and more. To these calls from Pan, Spock adds innumerable variations – the moral rules, the thoughts of what is dirty and what is clean, the social conventions and aspirations, the plans for long-term economic advantage. The dance is elaborate for many animals, but for us it is a jungle of conflict and, as always, the last things we can expect are coherence or consistency.

One thing I am fairly sure of – in this jungle, you cannot navigate to a fulfilling sex life using just the thoughts of Spock.

I was introduced to sex by a more experienced woman, and for a week or two, it was all rather constrained, with an uneasy sense that I was being watched. One summer night we were on cushions in the living room, naked, messing about. She had put on Pink Floyd's *Wish You Were Here* and to this day I love that album. Suddenly, out of nowhere, it was as if somebody turned a switch from A to B. It's impossible to describe except by the phrase 'letting go', though that phrase doesn't really say anything unless you already know what letting go feels like. Of course, I don't know what happens in a person's brain at the moment of letting go, but I can't help strongly suspecting that a major part is the inhibition of Spock, leaving the field free for Pan. This is not to say that Spock has gone completely quiet in a fulfilling sex life – evidently not, if we are wondering what our partner likes and thinking of adding more, or watching the elaborate cultural variations of courtship rituals. As always, we expect Spock to elaborate on themes from Pan, and the final result will be something they work out together. When it comes to sex, however, too much Spock may get in the way; not surprisingly, given the very long evolutionary history of courtship and mating, it may work best just to let Pan happen.

A jungle of conflicting instincts and thoughts does not sit easily with simple conceptions of ourselves. One person is different from another, the same person is different at different times; men are said to wonder 'what women really want' but the truth is that people want many, often conflicting things, and can switch with alarming speed from wanting one thing to wanting another.

In another book from the 1970s, *My Secret Garden*, the author Nancy Friday compiled sexual fantasies gathered from more than a hundred women – from conversations with personal acquaintances, taped interviews and letters sent in by women responding to an ad.[4] From Kinsey onwards, research has shown that most people have sexual fantasies, and in this book, Nancy Friday documented what the fantasies of women can be like. The back cover of my copy has a review from *Marriage Guidance*: 'Like *The Joy of Sex*, this book is for buying, sharing, enjoying and enrichment', and another, 'this book reminds us that sex can be wicked, raunchy, filthy and exciting'.

When I opened this book, I expected to find Pan let free in the unthreatening world of the imagination, and as we will see later, I think this is part of it . . . but the jungle is wilder than I could possibly have imagined. Being penetrated simultaneously by the eight tentacles of a huge black octopus . . . while Jesus is watching . . . who knew?! Nancy Friday says, 'If women are a mystery to men, they are even more mysterious to themselves and to one another.'[5]

Over and over again, these women end their interview or their letter by saying how grateful they are that at last they have revealed these thoughts to somebody, and at last don't feel that something must be wrong with them. We may not know exactly where sexual fantasies come from, but evidently they are something deep and fulfilling. As I say, the suggestion is that, sometimes, you just have to let Pan happen.

So what do women and men want from one another? It's a jungle . . . but there are themes. And though sex depends on letting Pan happen, as always we cannot just let Pan be our guide – in the twenty-first century, society needs rules to keep Pan in check. We need rules and laws on rape, harassment, decency. We are animals, and the sexual relations of other animals may be a guide to our own . . . but no other animal has our complexity, our flexibility and our capacity to reinvent ourselves. No doubt about it, it's a puzzle.

Many innate releasing mechanisms

In the last chapter, we saw the most common mating pattern in vertebrates. Large, aggressive males display and compete. Females assess and choose. There is an obvious attraction to the most muscular and aggressive male; this animal is likely to control the largest territory and provide best resources for raising young. From impala and hippopotami with their large harems, to songbirds competing to establish their territory, the strongest, most aggressive male achieves greatest success in attracting females. In return, he provides resources and defence, as we see from fish to greylag geese to the hunters and warriors of traditional human societies.

For the female, the advantage of a large, successful partner is not just in the resources and protection of today. The offspring of a large, successful partner are likely to be large and successful themselves. From insects to birds, females may favour these advantaged offspring; after mating, for example, the female domestic chick ejects more semen if it comes from a low-ranking male, while the female black grouse lays fewer eggs.

But this is the world of Pan, with competing considerations and competing IRMs. If there is the potential for resources to be shared, a female might also need a male who will favour her over rivals. In many species, that depends on individual pair bonding – not necessarily monogamous, but enabling a specific male and female to raise young together. In human societies, in fact, as in mammals generally, monogamy is the exception – though as Eibl-Eibesfeldt notes, 'Societies lacking marital relationships do not exist.'[6] Rivalry for the bond is also seen in many species, and in his list of human cultural universals, Donald Brown includes sexual jealousy. For the female Pan, there may be attraction to the strong, aggressive partner who offers protection and resources,[7] but the need for an extended bond means also a partner who will 'really love' her, and who will be 'a good father' over the many years it takes to raise a human family.[8]

We read Thomas Hardy's *Far from the Madding Crowd* (1874) or Margaret Mitchell's *Gone with the Wind* (1936) and know the dilemmas of choice between rival attractions. Should Bathsheba choose the glamorous, dangerous Sergeant Troy, who has already left one pregnant girlfriend to die in the poorhouse, or the manly and reliable Gabriel Oak, who will stay at her side till the crack of doom? We know humanity, and our hearts sink as the needle swings towards Troy. The same with Scarlett O'Hara and her choices of the dangerous Rhett Butler or the gentlemanly Ashley Wilkes. It is Pan, and two different voices call out, both demanding to be satisfied.

Each species is special, with its nature shaped by its own evolutionary forces, but often, a saga from another species can seem disconcertingly familiar. In some songbirds, for example, the female may first choose a male with a decent territory. Having mated with him, she also slips away and mates with another male, perhaps even more attractive. (As always, it is not just the female who is faithless – it takes two to tango.) The female returns to the nest and her first partner, who is none the wiser. The resulting young from both matings are raised together on the original territory, cared for by both female and first male – but sometimes, the female invests more in the chicks from her second partner, now long gone.[9] As I said, it makes evolutionary sense if this second partner was the more attractive of the two, because the offspring of an attractive male are likely to be attractive themselves – and in the next generation, to have a better chance of breeding.

These are not human beings, but it would take little imagination to turn their story into a very recognisable story of ourselves. Indeed, women's preference for physically attractive men applies more to short-term sexual encounters than to long-term relationships, and strengthens during the fertile phase of the menstrual cycle.[10]

Or another example from Lorenz.[11] As we saw in Chapter 6, mating provides an evolutionary challenge to an aggressive, territory-defending animal; somehow, the aggression released by the approach of a conspecific must be balanced against the urge to mate. In some cichlid fish, this balance plays out very differently for male and female. The male can chase the female round the tank, interspersing the chase with sexual signals, while the female flees, 'and at the same time make[s] use of every breathing-space to perform sexually motivated courtship movements'. But the other way round does not work at all – if the male is fearful, his 'sexuality is completely extinguished', while the female 'does not react to him sexually at all. She becomes a Brunhilde and attacks him the more ferociously'.

The human case is far more complex than the case of these cichlids but, again, something in the fish is recognisable. In her introduction to *My Secret Garden*, Jill Tweedie writes, 'Sadly for us, when men feel threatened sex goes out of the window.'[12] That rule may not be universal, but certainly, it is not unfamiliar. Very likely we simply have many IRMs of our own, sometimes resembling those of other species, all dancing together to produce our own sexual appetites and behaviour.

Troy and Oak

The Troy side of our human dilemma is certainly no surprise. If you ask my wife which male character most makes her knees melt, she answers without hesitation that it is Stringer Bell from the TV series *The Wire* (2002–8). (And my wife is certainly not alone in this – you may come across a little group of her female friends all agreeing about Stringer Bell.) It is not the actor Idris Elba, though Idris Elba in himself is fine – it is actually Stringer Bell, the cool, contained, lethal fixer who, no matter what needs to be done, will make sure that it is done to keep the drugs running. Behind the penetrating eyes of Stringer Bell, any thought could be forming. If you go with Stringer Bell and it turns out to be necessary, your body may later be found in a dumpster. It wasn't personal. It just had to be done.

In 'Revolution', the rather glorious final chapter of *The Female Eunuch*, Germaine Greer says, 'It would be genuine revolution if women would suddenly stop loving the victors in violent encounters.'[13] A revolution, perhaps, not about to happen soon . . . and a comment strongly reminiscent of Lorenz on the Egyptian goose, who 'watches all the fights of her mate with the interest of a boxing referee [and] if he comes off the worse, she is always ready to go over with flying colours to the side of the winner'.[14]

Or we could think of E.L. James's 2011 novel *Fifty Shades of Grey*, with its themes of dominance and submission. According to Wikipedia, *Fifty Shades of Grey* has been translated into 52 languages and set a UK record as the fastest-selling paperback of all time. By 2017, the book and its sequel had sold 50 million copies worldwide. I am not sure how many of the readers were women, but the popular belief is most of them.

I have said that, in *My Secret Garden*, there is everything you can imagine – and a lot you never would. But if there is one thing that is most common, it is submission, rape and anonymity. Often the aggressor is unknown, often coming from behind, often with several men at once, often with a claim that 'I feel I might really enjoy this – if I were forced to do it'. Subsequent, more formal research very much confirms this picture, with force and submission appearing among the most common themes of women's fantasies.[15] As Nancy Friday is at pains to point out, usually (not always) the responder is clear that this is only fantasy – she would never actually want it. As I say, it seems likely to me that this is an element of Pan set free in a place she can do no actual harm. In the jungle are calls to which we would never really respond – but we can still hear them.

I am not sure how far back in our distant past these calls can originate. One thing I did not expect at all in *My Secret Garden* was the frequent appearance of many men at once, with the woman forced to have sex with one after the other. As far as I know, this kind of sexual activity is not at all typical of *Homo sapiens*, but it is very typical of our near relations the chimpanzees, where males all mate in series with a receptive female. Is there a behaviour pattern here from some distant common ancestor, long overlaid with our own, pair-bonded mating strategy, but still buried somewhere in our minds? Of course, we have no idea but, once again, it's a jungle in the crazy sexual world of the human being.

All of that said . . . in a long life I have met a number of men who were especially popular with women, but perhaps the most popular of all was about as far as it is possible to be from Sergeant Troy. He was a small, frail man with a polite, almost deferential manner. I remember the first time I met him, walking along a street in north Oxford, and I was disposed to disregard him altogether, he seemed so meek. Then we became friends and I became very familiar with the fact that pretty much every woman in college was all over him. I'm not sure it did them much good, as he was already married to a rather glamorous film actress. He was something else, and the source of his success was obvious. He simply loved women. He gazed admiringly into their eyes. He touched their elbows (perhaps you have guessed that he wasn't English). He wanted to hear what they had to say, but he also had a lot to say of his own. He suggested they go out to the movies. He just made women feel great. Another character from the jungle.

Troy gives a shiver of attraction but, evidently, the need for reliable, long-term pair bonding is also dominant in *Homo sapiens*. We are one of the species in which male and female both contribute to care for the offspring, in our case extended over many years, and much has been suggested about specific human adaptations for promoting long-term bonding. Very likely, all those bonding mechanisms we saw in Chapter 6 – the laughter together, the dancing together, the favours given and received, the shared goals and challenges, the shared memories – act to cement the bond of the human couple. Very likely, too, we are like prairie voles in the specific impact of mating, with its associated release of hormones including oxytocin and vasopressin; unlike most other primates, the human female is sexually receptive even when she is not ovulating, allowing frequent, repeated sexual activity. Our bodily features are quite special and, often, attract a spectacular degree of attention. A woman's breasts are much larger and rounder than the corresponding breasts of other primates, and exert a well-known, magnetic attraction on the eyes of men. The man's penis is also much larger than the corresponding organ of other apes[16] – the erection of the mighty gorilla, for example, amounting to a modest 5cm. In the fantasies of *My Secret Garden*, the penis is almost always exceptionally large, and woman after woman gives a report like this one from Una:

> I myself am so unconscious of looking at men, at glancing at their crotches as they approach me on the street, that I can be thinking of what to buy for dinner while my mind is speculating on just what a man has done to himself to achieve a particularly interesting arrangement of his genitalia.[17]

Or Myrna, 'Naturally, I look. Doesn't everyone?' Nancy Friday herself concluded, 'They all look.'

Though everybody agrees that 'size doesn't matter' when it comes to function, very likely it does matter when it comes to the penis simply as a releasing stimulus, attracting sexual attention like the releasers of many animals' sexual IRMs. The fantasies of enormous organs in *My Secret Garden* are strongly reminiscent of the ethological concept of the super-stimulus, as we saw in Chapter 2: the sight of an egg that has rolled out of the nest strongly drives the parent oystercatcher to roll it back in, and when the experimenter substitutes a ridiculously large fake egg, the determination to roll just gets stronger.

In a classic of the 1960s, *The Naked Ape*, Desmond Morris speculated that the expanded breasts of the human female evolved as a specific impetus to pair bonding. In all other primates, he pointed out, the male mounts the female from the rear. In many species, specific signals release this mounting, such as a bright colour patch developed only when the female is receptive, and visible as she is approached from behind. In humans, Desmond Morris thought, it became important to mate from the front, encouraging focus on the specific identity of the partner. He speculated that the breasts became large and round, reminiscent of the attractive buttocks of an evolutionary ancestor – but now transferred to the female's front, encouraging mating also to be transferred to this orientation.

I have no idea whether this thought from Desmond Morris has any truth to it, but what I am certain of is that, in humans, the strongest signals of sexual attraction come from the front. In particular, they come from the eyes. We can take it from Carly Simon, celebrating the real thing in her song 'Look Me in the Eyes'.[18]

Or Isabel Allende, whose lovers in *The House of the Spirits* (1982) take up

> yoga exercises and meditations . . . as a couple, seated face to face in complete relaxation, staring into each other's eyes and murmuring Sanskrit words that could send them all the way to nirvana but that generally had the opposite effect, and they would wind up . . . desperately making love.[19]

I remember the precise moment when, as a very shy young student, I discovered this for myself. Somewhere in Oxford, I was at a concert performance of Mozart's *The Magic Flute*. In mid aria, from across the hall, I saw that a woman was gazing at me. I hesitated, gazed back. I was astounded. My heart was pounding. My penis stood up. Feel free to laugh – I was young, I was shy, I didn't know this! Shortly after, it happened again – watching a cricket match! – with the same results.

Between men and women, at least in my experience, the strongest stimulus to sexual attraction is not the large breasts or the large penis, it is the gaze of mutual interest. We speak of 'feeling a connection', and to a large extent, I suspect, this almost spiritual sense is little more than gazing into one another's eyes. I suspect that sometimes, when a woman feels that a man is looking at her breasts and she is treated as a 'sex object', the problem is not simply that he is showing sexual interest. It may just be this, if the interest in itself is unwelcome, but perhaps at least as important, he is showing his interest in a crass way. The breasts feel like an 'object', while the eyes feel like a person. When he looks into her eyes, there may still be sexual attraction, but now it feels personal – and much more likely to be welcome.

The rival attractions of Troy and Oak raise an intriguing question over the fantasies in *My Secret Garden*. As I said, when

I opened the book I expected to find the many sides of a female Pan, with many conflicting urges coming to the surface of the imagination. The Stringer Bell side comes up over and over, as in this quote from Sadie:

> I am being raped by one man or a group of men, while many of them watch the others 'abuse' me. My attackers are always very handsome – dark hair, muscular, sexually well endowed – and brutal, in that they take what they want and the hell with what I want . . . or pretend I want. (I'm after what they are, really.)[20]

But what of the gentle, caring partner who nurtures and shares? This partner is all but absent from the whole book and, sometimes, fantasy and reality are put into direct opposition. Heather describes fantasies of a previous, rather cruel partner, but concludes, 'As it is, I am going to marry my boy friend. He will make a good husband and father, but I am afraid that I may go through the rest of my life feeling something is missing.'[21] While Rose Ann makes love to her husband, she fantasises about being stretched, 'one brutal man on each limb', and wonders, 'how can I possibly tell my husband, whom I love, that I am dreaming that the most atrocious things are being done to my body while he is being so loving to me?'[22]

It seems that much good sexual fantasy needs the fear side of the equation, except . . . in one section, there is a whole list of fantasies about sex with other women, often from women who do not at all want actual female partners . . . and now, caring and tenderness are all over the page. Make of that what you will . . . meanwhile, it seems that even the most caring husband might be wise, once in a while, to mix at least a dash of Troy in with the regular diet of Oak.

Men are complex too

Especially in the context of gender feminism, it is sometimes proposed that, for men, all sex is a matter of power. It is there to preserve the patriarchy. What men want is not a partner, but a slave. Gender feminists can certainly be pretty negative about sex, in a way that, once again, is classic Spock, running with his (in this case, her) idea to a place that is far removed from any recognisable reality. Here is Catharine MacKinnon:

> [Male] force is sex . . . Hostility and contempt, or arousal of master to slave . . . Violation, conventionally through penetration and intercourse, defines the paradigmatic sexual encounter . . . what they want is women bound, women battered, women tortured, women humiliated, women degraded and defiled, women killed.[23]

As usual, in the complex, conflicted world of the human mind, it is easy to see that this idea has some truth to it. In Chapter 8, we saw the potential men have for callous, uncontrolled brutality. Men can be brutal to an enemy, or simply to take what they want from somebody weaker, and certainly they can be brutal to women. To estimate the true rate of rape worldwide is impossible, given unknown levels of under-reporting and under-recording, but whatever the number is, it is immense. As one example, a recent count of reported rapes in the United States gives a rate of around one rape every one to two minutes.[24] It would be similarly impossible to know true rates of domestic violence, along with many other forms of men's violence against women.

As I mentioned earlier, when Eibl-Eibesfeldt asked the Yanomami why they waged war, one of the primary reasons given was to capture women; a theme in early warfare already noted by Darwin. The concept of 'rape and pillage' strongly suggests that, when a group of men overwhelms an enemy settlement, violation of the women is a strong urge. Of course, we do not know how much of this is built into male human IRMs, but very likely, some of it is. Research shows that dominance appears commonly in the sexual fantasies of men as well as women. Sexual coercion is certainly known in other primate species, including orangutans and chimpanzees.[25]

At the same time, there are animals with strong inhibition against male attack on females; Lorenz gives the example of male hamsters, much larger and stronger than females, but so strongly inhibited from fighting back that, if they are kept together after the mating season, the male may literally allow himself to be bitten to death.[26] At the least we may say that, in our species, any such inhibitions are not at all sufficient to prevent substantial, sometimes lethal violence.

But this is Pan, with all his conflicting urges and IRMs, and to me it seems very far-fetched to imagine that this is the whole story – or even, in our daily sex lives, a very major character. Men, like women, want many things from their sexual partners, and what they want varies wildly from person to person and from time to time. If I were to put my finger on a single, most common theme of male fantasy, it would not be rape and domination. Instead it would be exactly what you would expect from any displaying songbird, singing his heart out in the hope that females will respond. Indeed, it is the familiar fantasy (and often reality) of a man on a stage with a guitar. In his song 'I'm a Real Man', tongue firmly in cheek, John Hiatt boasts of the many women (it is really rather a lot of women) who would pay to spring him from jail. In Robert Cray's 'Fantasized', a waitress from a local café turns up at his door, irresistibly drawn by the power of his flirtation. This one is actually called 'Fantasized'! Links to both are in the Notes.[27]

Like many women in *My Secret Garden*, dreaming of arousing uncontrollable lust, these men just want to be admired. It is not 'I forced her into it.' It is, 'She fancied me so much she couldn't help herself.' Research on men's sexual fantasies shows that the pleasure of the partner can be an even stronger theme than dominance and a sense of power.[28]

And much more. The intense commitment to childcare, so strong in men as well as women, makes it seem certain that men, like women, are looking for the woman who will bear and raise strong children. Across cultures, men prefer women with youth, health and social resources.[29] Beyond childcare, very plausibly, there may be search for the partner who will help and support through all the vicissitudes of life. In *The Caine Mutiny* (1951), Herman Wouk's troubled protagonist seeks out the woman he loves but intends to leave because he feels she is not good enough for him; this intention notwithstanding, he wants to tell her that he will be court martialled. Rather movingly, he has no understanding of his own need for life's closest friend: 'He still had not the slightest understanding of why he had really come . . . He had no way of recognizing the very common impulse of a husband to talk things over with his wife.'[30]

And once bonds of different sorts are formed, it seems certain that there are IRMs to defend and protect, just as a male baboon will protect his grooming partner from unwanted sexual advances by other males. For many years, I was amused by my reactions to a senior administrator in Cambridge, a powerful and much feared woman who held the purse strings, terrified alpha male professors with her instantaneous grasp of the facts and, on several occasions, helped me to navigate through the thorns of scientific politics. I viewed her with the strongest admiration and respect, yet she was extremely short, just like my little sister, and even as I listened to and thanked her, and always took her advice, I knew that I wanted to put my arm round her and protect her. From what?! If anybody did not need my protection, it was her.

Intriguing results show just how much of Pan there is in regulating these relations. In rats, for example, aggression is inhibited by a chemical found in the tears from a female eye, and in humans it is just the same.[31] When men sniff female tears – quite unaware of any odour – their testosterone is reduced, along with their aggression, while brain imaging shows reduced activity in the amygdala.

For more complex cases, the causal mechanism is unknown. We cannot possibly know that a man's brain contains IRMs for Wouk's 'common impulse of a husband to talk things over with his wife', but it is perfectly plausible that it may contain just that – some group of neurons, like the small groups of neurons that control attack and courtship in a mouse, that respond to a social threat, plus the presence of the partner, and perhaps the hormonal aspects of both, and release a desire to talk things over. For now this is speculation, though to me it is plausible that one day we will know the neurons, the hormones, the genes and how it is that Spock turns a forthcoming court martial into Pan's signal of a threat.

Spock's rules: No means no

Critical though they are in managing our individual sexual relationships, Pan's promptings on their own cannot run society. As we have seen in previous chapters, we need Spock to provide rules and laws that can balance the conflicting impulses of millions of individuals. Most obviously, we have the prohibition of rape, and the maxim 'No means no.' To run our society, we need rules that will allow women to be safe, confident and free to do as they please.

To understand 'No means no,' we need to remember what this phrase actually is. It is wrong to interpret it as a belief about the complex real world, where there are many varieties of 'no'. In Mikhail Sholokhov's *And Quiet Flows the Don* (1940), there is a shattering scene of dull, brutal manhood, as one soldier after another takes turns with the destroyed, inert body of a kitchen maid. But different from this there is Scarlett O'Hara in *Gone with the Wind*, tormenting her drunken husband Rhett Butler with the love she has always felt for the insipid Ashley Wilkes; at last chasing her to the bottom of the stairs, he declares that 'this is one night when there are only going to be two in my bed', seizes her and carries her, resisting and afraid, upstairs to the bedroom . . . cut . . . and she wakes up singing.[32] (Too late to save their marriage, as it turns out.)

In today's terms this is rape, and Rhett Butler might fairly be prosecuted . . . but Scarlett's singing heart the next morning is not unreal, either. In *Who Stole Feminism?*, Christina Hoff Sommers quotes a survey of more than 400 female fans of *Gone with the Wind*. The great majority said that this was not rape (the survey was published in the 1980s, and very likely the figures would be different today) but erotically exciting 'rough sex' – Sommers herself suggests that Butler does not 'rape' but 'ravishes'. Dissatisfied with her gentle husband, Maria in *My Secret Garden* tells us, 'Often, lately, I have resisted having sexual intercourse with my husband when he wants it (which is only about once a month anyway) so that he will have to force me to have it with him, in the hope that he might sort of rape me.' Maria finishes, rather forlornly, with, 'So far, though, he has not done so.'[33]

'No means no' could be generalised even beyond sexual contact. I once read an article about Barack Obama, recounting how he had to ask Michelle out several times before she agreed. Sometimes 'no' means '. . . but you can ask me again tomorrow'.

The jungle of all these different cases is no real surprise. Given our vertebrate heritage, a pursuing male and a more cautious, selective female are very likely built into our genes, with both the tendency of the male to pursue and the tendency of the female (sometimes) to be attracted by that pursuit. The flourishing rape scenes in *My Secret Garden*, the spectacular sales of *Fifty Shades of Grey*, my wife's views on Stringer Bell, the reactions to the 'ravishment' of Scarlett O'Hara all suggest this is true.

But in the complex world of human communication, 'No means no' is not supposed to be a complete description of all that goes on in the conflicted minds of Scarlett O'Hara or Maria from *My Secret Garden*. Like 'Thou shalt not kill', it is a rule specifying how we must act – that sex without consent is not permitted. Like all such rules it simplifies reality, but is indispensable for life in a huge, anonymous community, where every woman on the street must be protected from the potential violence of every man. We know the world is complex, but this phrase captures the most important truth that could be captured in a simple, enforceable rule. It is an essential cultural agreement over how we are to behave.

There is another question about rape where, again, we need to remember exactly what things mean. Sometimes it is felt that a woman is 'responsible' for what happened to her – perhaps because she was dressed 'provocatively', or alone in the wrong bar at the wrong time, or unconscious after too much alcohol. These thoughts may feel right, but if we accept my analysis of 'responsibility' in Chapter 10, they are not. Again, we should remember the distinction between moral or legal responsibility and cause, often confused in our minds but quite different. When something goes wrong, to say that somebody is morally or legally responsible means precisely that we hold them responsible – we have rules of behaviour, and the person broke those rules. To ask what caused something is much wider and, as we considered earlier, the causes can always be traced back indefinitely far into the past. Much in a woman's choices may contribute to the chain of causes that led to her rape – if she had made different choices, the rape would never have happened – but that does not mean that she is 'responsible' in the moral or legal sense. She has violated no rule by the way she dressed or by where she chose to be – quite the contrary, we value a world in which she is free to dress and to go as she pleases. Instead, it is the man who has broken one of the most important rules in our society and, having done so, is to be blamed, scorned and punished.

A harder case

Along with rape, women expect protection from many other kinds of unwanted sexual attention. Again we need rules . . . but in our conflicted sexual world, such rules are not easy to define. A few years ago, over a long lunch, I argued with a female friend over whether women 'want' sexual attention at work. We went back and forth for a good period with examples and counterexamples. She spoke of the sense of shrinking back at an unwanted touch from a senior colleague. I spoke of shared flirting and the sense of adventure – and for that matter, the 50-year relationships that sometimes follow. Eventually we reached the splendidly obvious, useless but undoubtedly correct conclusion. Women want sexual attention at work . . . when they want it. For a (much) funnier version of this same conclusion, you could try Michelle Wolf's take on Me Too in her 2023 show *It's Great to Be Here*.

Once again, the difficulty with going further is that the balance of conflicting considerations varies wildly across person and time. On the one hand, there are many men willing to use their power, either organisational or physical, to impose unwanted attentions. The abuse can be staggering, sometimes heartbreaking. From the BBC:

'Dead soldier suffered relentless sexual harassment – Army report.' Ten days before Christmas, 19-year-old Jaysley Beck killed herself after months of harassment by her immediate superior. The army investigation found that 'in October 2021 her boss sent her more than 1,000 WhatsApp messages and voicemails. The following month this increased to more than 3,500.'

'Female surgeons sexually assaulted while operating.' Among many other reports, a female surgeon describes how, as a young surgical assistant, she had a senior male surgeon wipe his brow in her cleavage; offered a towel, he said, 'This is much more fun.'[34]

On the other hand, there is the appeal – sometimes the substantial appeal – of attention from an attractive partner, and the question of how far it may go. The fact is that the great majority of my married friends met at work, with varying degrees of separation in status – and in most cases, are still married decades later. In my world, this happens a great deal more often than surgeons burying their face in a colleague's cleavage. But both happen.

The plus and the minus another way: a few years ago, in the London Underground, signs appeared saying, 'Staring is harassment too.' Of course, I was reminded of my experience in the opera and, even more strongly, of an Israeli friend from student days. She had been an officer – I think a major – in the Israeli army, and as this was right after the October War of 1973, it is a good bet she had seen military action. In any case, she was not somebody you would ever mess with, and once after a brief trip home, she described how Israeli men, unlike the shy English, stared at her in the street, and she thought, *Thank God – I'm a girl again.* Of course, to another woman, perhaps less self-assured, or younger, or from a different culture, this staring would be felt as unwelcome, intrusive or dangerous.

Again, one size does not fit all, and in the Underground, a better sign might perhaps have read, 'Unwelcome staring is harassment'.

Again, too, the weights of our rival impulses depend heavily on the situation. I think I have never had a woman play at catching my eye in the London Underground – where there is perhaps a feeling of rats trapped in a buried world – while it has very often happened on trains in the normal, above-ground world or in restaurants. Perhaps the sign should have read, 'Staring in the Underground is harassment . . . but above ground it may be OK.'

We are balancing the rival considerations of excitement and opportunity on the one hand, a right to be left alone on the other, and in the decades of my life, the pendulum has swung strongly in the direction of the right to be left alone. Simply when you walk up the street, the cultural variations are obvious. In the 1970s, I expected that passing women would be looking into my eyes and checking me out – in a trip to the US in the 1990s, I noticed that it had stopped . . . then fast forward to 2011, when I was trapped for a few days in Tel Aviv waiting for the Iceland ash cloud to disappear, and suddenly it was like being young again, with every passing young woman giving a frank, curious look into my eyes.

With all these variations over time and place, it is impossible for me to imagine where the pendulum 'should' be. With everything gained, something else is lost. My sense of 1970s feminism was that women felt powerful, free, proud, ready to grab what they wanted from life. Fifty years later, the mood of the culture is that all of us – men as well as women – are potential victims who need society's protection. The need is real enough, as the story of Jaysley Beck and many others shows. It is also just a part of the truth; and there must be costs to a social context that encourages fear over individual self-confidence.

Many years ago, in a quiet spot close to my work, a man had been seen hanging around apparently doing nothing. Several of my female colleagues felt threatened, and when we reported this man to the police, they confirmed that this location could indeed be a danger spot (though as far as I know, this particular individual never did anything further). The threat was real, and potentially serious, but I will always value the response of my wife when I told her. It was, 'But is he just a harmless flasher?!' Obviously, flashers are not at all harmless, and sometimes very dangerous, but it is a lucky woman whose first thought – perhaps even an unwise one – is that it takes more than a flasher to mess up her day. (Another dear, rather small, friend of mine used to boast of how once, in the streets of Los Angeles, she and a female friend fought off a pair of very large, very determined carjackers, with a kick to the testicles playing a starring and especially satisfying role. In her case, even carjackers were not allowed to mess up her day. She is a crazy woman, and few of us, perhaps, can aspire to her level of self-reliance – but how great to be this crazy.)

At work, we can easily appreciate the advantages of suppressing the animal, and allowing all an equal chance to be a productive operational unit. It seems certain that, at least in part, the increased sexual hesitancy of the twenty-first century is a response to this desire for fairness across the many roles of society. Unlike many gender feminists, I do not for a minute think that a man's sexual response to a woman necessarily demeans her, or diminishes his respect for her brilliance and achievements – in my mind, the attraction and the respect are just two different things, just as likely to support as to inhibit one another – but still, sex can sit uneasily with fairness, and once again, it's the conflicted jungle. Pan can be a strongly resented nuisance, and sometimes heartbreakingly damaging, but suppression also has its losses.

I am often reminded of my mother, who was extremely proud of her ability, with just a look and a smile, to make men do what she wanted. For a few years she drove a flashy Volvo sports car, eye-catching pearly white, and as she had little respect for speed limits, she was periodically stopped by the police. As long as it was a man, she was always confident that, after a brief exchange of pleasantries, she would get off with a warning – once, she was extremely offended when the impertinent policeman was actually so crass as to fine her. Slightly reminiscent of my mother was one of the secretaries at my work many years ago, who also drove a somewhat flashy car. On the front she had a bumper sticker that read, 'If you think I'm sexy, flash your lights.' On the back was another that read, 'I said your lights!' Both these women seemed extremely happy in their skin, and not at all diminished; and certainly my mother, often to quite an infuriating extent, was always more than willing to begin issuing instructions to any man, woman or child standing in her vicinity.

I have my own rule for navigating the complexities of sexual attraction. It is deeply unenforceable, it is hard to be sure you are obeying it, and it may strike the reader is hilariously British, but it has the merit of being actually believable. It is: 'Don't be an arsehole.' Surely, in all its complexity and variety, this is the rule that we try to teach our children. It would take a very long poster on the London Underground to specify everything that it means, with all its dependence on time, place, person and circumstance. Luckily, we have all the complexity of our inner Spock and Pan to tell us.

Who is in charge?

Does an attraction to Troy mean that it is 'natural' for a woman to be submissive to her partner? Though it is true that, in vertebrates, the male is usually more powerful, this is surprisingly unpredictive of relations between partners.[35] Sometimes it is just as you would expect. Lorenz gives the example of the emerald lizard, in which even a much larger, stronger female instantly submits to any male, but in other species, many other things happen. Sometimes it is the opposite, as we saw earlier in the chapter with the female hamster biting her larger consort to death. Even closely related species can behave quite differently. In the greenfinch, the female dominates the male during the breeding season, but positions are reversed for the rest of the year. In the bullfinch, the female often pecks at the male, who never pecks back. She appears dominant, but Lorenz again reminds us how human these exchanges can appear (and perhaps his interpretation is a little too human):

> When a male bullfinch is pecked by his wife, he in no way assumes the submissive attitude but, on the contrary, he shows sexual self-display and tenderness. Thus he is not pushed by the pecking of his wife into a subordinate position, but, on the contrary, his passive behaviour, the manner in which he accepts his wife's attacks without becoming aggressive and without letting himself be put out of a sexual mood, has an 'impressive' effect – apparently not only on the human observer.[36]

When it comes to humans, the story again is that we can see almost anything, with extreme variations over person, place and time. In many cultures, certainly, the wife is essentially (or even formally) the property of the husband, and is expected to obey his commands. But in many other cultures, including our own, this is not at all true, and certainly in daily life, I have witnessed bullfinch considerably more often than emerald lizard. In an alternative version of *The Wire*, it is easy to imagine that, after a hard day in the drug wars, Stringer Bell goes home to his wife and is told in no uncertain terms that he has left his gun where the children might find it. Everyday life is surely enough to tell us that, in both males and females of our species, there are strong tendencies for both aggression and submission.

Looking back at the distributions in Chapter 12, distributions for the two sexes are likely not the same, but strongly overlapping as well as strongly varying over time and releasing conditions. Add this to the strong influences of Spock and his cultural expectations, and you have the spectacular mixtures of human relationships.

We should also think again how IRMs that dominate in one context become quite irrelevant in another. In *My Secret Garden*, Nathalie completes her description of dominance/submission fantasies (in both directions) with: 'since I've been involved in Women's Liberation . . . I am sure there are other women like me, who having emerged from being under male domination, crave to return to it in bed.'[37] I am also sure that there are women who don't, but Nathalie reminds us just how strongly we resemble any other animal, with needs and behaviour patterns dependent on the exact time and circumstances.

Where does this leave us with rules for a good marriage? The benefit of a clear authority ranking relationship – like the benefit of a clear pecking order – is that it maintains harmony. Conflict is minimised when everybody accepts who is in charge, but at the price of fairness, and it does not take today's Taliban-ruled Afghanistan to show what happens with severe asymmetries of power between men and women. In modern Western culture – especially American – I have a sense of the opposite; in modern Hollywood dramas, it is implied that women are somewhat letting the sex down if they are not slamming a door and telling a man he is worthless. This is great for fairness, less great for harmony.

My own preference is to avoid the authority ranking mode altogether. A relationship much closer to Fiske's equality matching is well captured in this exchange from Rudyard Kipling's short story 'Friendly Brook' (1914). It is an exchange between two men, but to me that is irrelevant. The men are contemplating a hedge that has grown far out of shape and needs to be laid. (I think this makes sense even if you have no idea what it means to lay a hedge – the important point is, you need to know what you're doing.) Kipling says:

> Jabez rubbed his wet handbill on his wetter coat-sleeve. 'She ain't a hedge. She's all manner o' trees. We'll just about have to –' He paused, as professional etiquette required.
>
> 'Just about have to side her up an' see what she'll bear. But hadn't we best –?' Jesse paused in his turn, both men being artists and equals.[38]

As the saying goes, this is not rocket science. It's kind of obvious how equals should treat one another.

CHAPTER 14

Weights

The kibbutz experiment

Imagine the perfect, rather dystopian experiment to test the predictions of gender feminism. A new community is set up committed to the equal treatment of boys and girls. From birth, these new children are brought up with no idea of sex differences, and no exposure to the world outside the experiment. Thirty years after the experiment begins, we see what sort of people it has produced. In particular, we see whether the abolition of sexual differentiation in teaching, expectations, experience and power has produced men and women who are psychologically indistinguishable.

This experiment will never be done, but in the kibbutz movement of what is now Israel, something as close as we could ever get was in fact done. In 1950–1, the anthropologist Melford Spiro carried out his first field study of a kibbutz he called Kiryat Yedidim, then about 30 years old. Describing himself as a committed cultural determinist, he wished to document the 'revolution in human nature' that he expected from the radical new social structure created by the kibbutz movement. In 1975, he visited again, producing the first edition of his book *Gender and Culture*, and then in 1995 released a new edition documenting how the changes he had seen in 1975 had been amplified over a further 20 years.[1]

As Spiro notes, he never particularly intended to focus on the relations of men and women. He certainly never expected what he found. The kibbutz movement did not produce a revolution in human nature. Instead, against all the ideology, practice and reluctance of the original founders, human nature produced a counterrevolution in the kibbutz, and as Spiro puts it, a counter-revolution in himself.

As described by Spiro, from its foundation in 1921, the social structure of Kiryat Yedidim followed the common principles of the kibbutz movement. Heavily influenced by nineteenth-century socialism, the founders believed that cooperation and brotherhood should be the ruling principles of human society; that land and goods should be publicly owned and shared on a principle of need; that everybody should work equally, in particular on the land; that decisions should be communal, based on pure democracy.

The founders had a specific belief on the relations of men and women. Women, they believed, were kept 'in shackles' by the need to bear and raise children, leading inevitably to political and economic dependence. At the same time, they believed, family life left children under the 'patriarchal authority' of the father. To realise their utopia, they produced a radical dissolution of the family. Children were seen not as the children of their parents, but as the children of the kibbutz. They were raised not in family units but in age-graded children's houses, under the supervision of childcare specialists. The arrangement left both mother and father free for work on the land, and free to realise the true independence and equality that this new social order would create. Though parents were permitted to spend time with their children, this time was strictly limited (on weekdays, two hours after the end of the working day).

In the children's houses, boys and girls were not differentiated. As Spiro summarises:

> In each house boys and girls played, slept, ate, and showered together, and (during their toilet training) sat on their training pots together. As far as possible, the socialization of both sexes was the same. Moreover, boys and girls shared the same toys, and all play and games taught them by the nursery teachers were sexually integrated and undifferentiated. The same pattern characterized their other learning experiences. Boys and girls alike were inculcated with the same values concerning the importance of agriculture and labor. They worked together in the 'children's farm,' comprising a vegetable garden, some sheep, and a poultry run . . . The children's experiences in their two-hour visit with their parents were little different. Except for the fact that babies were nursed by mothers, mothers and fathers displayed one parental role, rather than differentiated 'paternal' and 'maternal' roles. Not surprisingly, therefore, when in 1951 we elicited descriptions of the socialization roles of their parents from these children, they described only minor differences between father and mother.[2]

In the earliest days of the kibbutz movement, Spiro records that even breastfeeding was regarded as a community, not individual, activity. Mothers nursed their babies in a group, and babies were weighed to ensure that all ate sufficiently. If a mother's baby had already gained enough weight, she was expected to move to a new baby that had not eaten so much.

Spiro is also clear how firmly the kibbutz founders were committed to these principles. Though physically weaker, women worked long hours on the land to match the production of the men. Marriage ceremonies were abolished; instead, the married couple simply received a joint room, and in daily life, avoided public shows of affection and referred to one another as 'comrade'. Commitment to the children's houses was total, with a strong belief that family childcare could not be reconciled with the ideals of the kibbutz movement. In 1950, Spiro says, he was often told, 'The kibbutz may undergo many changes, and still remain a kibbutz, but if collective education were to be abolished, that would be the end of the kibbutz.'[3]

Of course, as a test of the principles of gender feminism, this experiment was not perfect. Undoubtedly, children growing up in the kibbutz would also have had some knowledge of the world outside, where conditions were very different. From the early days, furthermore, a degree of role differentiation re-established itself; the children's houses, for example, were soon staffed almost entirely by women, in part because their weaker bodies made them less suitable to the hard work of the land. Still, it is hard to imagine that anybody could ever try harder than the kibbutz founders to eliminate the psychological and social differences of men and women.

Twenty-five years after his first investigation, Spiro returned to see how the new generation, raised in their children's houses, had developed into adults. What he found has profound implications, not just for the Pan and Spock of human nature, but for freedom of choice, the essential difference between individuals and distributions and, above all, for concepts of equality and value.

The feminine revolution: Family

By 1975, the children from Spiro's 1950–1 field study had grown up. In Hebrew, the word for a person born in Israel is *sabra*, but, as Spiro shows, it was the female *sabras* who had driven the changes he saw and accordingly, in his book, he uses *sabras* just for the women. As he puts it, the kibbutz pioneers had instituted a feminist revolution. The *sabras* grew up and responded with their own revolution – this time, a feminine revolution, already firmly established by 1975, and even more so by the time of Spiro's 1995 retrospective.

At the heart of this revolution was a slow revolution in child-care. The ideology of the kibbutz was committed to collective childcare and the abolition of the family, but as they grew up and had children of their own, the next generation of *sabra* mothers did not like it. Against strong opposition, a series of changes took place, more rapidly in some kibbutzim than others, and slowly taken up in Kiryat Yedidim itself. Initially, the mother's workday was reduced so that she could spend more time with the child at the end of the day. Later, this was followed by a system called 'the hour of love' – now, a mother could take first half an hour and then a full hour out from the middle of the day, initially if her baby was very young, then later extended to apply to older toddlers. Finally, there was the most radical change, which in Kiryat Yedidim was not established until the 1990s. Sleeping in the children's houses was abolished. Instead, at last, children were allowed to sleep – and perhaps even more important, to wake up – with their parents in the family home.

As Spiro emphasises, it was not the children who insisted on this change. In the early days, he says, children might well enjoy their time with their parents but, in most cases, were also quite happy to return to the communal house to sleep. This house, after all, was the home they knew. Neither was it the men. Men, on the whole, were happy to support home sleeping, but this was because their wives insisted on it. It was the *sabra* women who demanded change, and reading Spiro's interviews, and the interviews he quotes from other kibbutzim, their reasons are compelling – coming as they do from women raised in this extreme natural experiment in cultural determinism. They say: 'For me it is clear that the greatest loss for me as a mother was that I did not have the privilege of raising my children in closer proximity to me . . . We created a beautiful framework, but sometimes it works against the laws of nature'; 'There is a conflict here between ideology and emotions . . . as a mother, I am dying to have my baby sleep at home, and to see him wake up in the morning'; 'When the mother returns to her apartment after she has put her child to bed, she feels that something is lacking. Something is lacking, too, when she cannot awaken her child in the morning and watch him open his eyes . . . She wants the child with her because that is how her maternal feelings are fulfilled.'[4]

These comments come from the *sabra* mothers but, rather tellingly, Spiro notes that the same feelings now emerged, possibly at redoubled intensity, in the grandmothers, those same women who as mothers themselves had stayed firmly committed to the founding principles of the kibbutz. One kibbutz woman put it like this: 'These grandmothers are expressing all those natural feelings which they had suppressed when they were mothers . . . Now that their daughters and granddaughters are saying, "Nonsense, my children are mine," these pioneers agree with them.'[5]

In the account that the *sabras* give for their feelings, over and over again they explain that these feelings are natural, biological, like the feelings of other animals. The women grew up with the chance to be finally emancipated from the shackles of family life, and the firm expectation that this chance would be taken. As it turned out, they simply didn't want it. In their small apartments, family sleeping meant 'crowding . . . admitted restrictions on their sex lives . . . curtailment of their extra-domestic activities', but interviewed in 1994, 'not one woman expressed regret for having made the transition'.[6]

The feminine revolution: Work

As for family life, so for work. At the start of the movement, as I have said, the expectation was that men and women would share equally in the work of the kibbutz but, over the decades, this expectation was gradually eroded. Hard work on the land became a male speciality; 'service' jobs, including childcare, cooking and education, became predominantly female. The women comment that agricultural work was too physically demanding, or that they were not interested, or especially that it kept them too far from their children. Spiro puts it like this:

> Today this obsessive concern to prove their worth as women, by demonstrating that they are as good as any man – in the things that men do – is dead. For the older sabras, it has become a historical memory; for the younger ones it is merely another of those 'quaint' ideas that the pioneering generation had dreamed up.[7]

He cites the case of 'Ruth', 'a rate-busting field worker when I knew her in 1950 . . . Angular and strident in her twenties, she had become in her forties a jolly and warm woman'. Perceiving Spiro's amazement, Ruth explains it had been 'crazy' to believe that agricultural work was important while work in the kitchens was not. 'To feed cows or chickens was "farm work," hence "good," but to feed people was "service work," hence "bad." One day I decided that's just crazy . . . So I went to work in the kitchen, and I loved it.'[8]

Again, the *sabras* are quite clear on the reasons for these work choices. The say that 'women . . . are most fulfilled by working with and helping other people, while men are most fulfilled when working with machinery and in tasks which give them a sense of power and domination'.[9] Or, 'why should I work in a branch that does not interest me? I am interested in . . . education. I like to work with human beings.'[10]

Over and over again, they say that it is the family that matters most to them. Spiro cites studies in several kibbutzim, showing that, for the majority of men, work was rated as more important than family, while for the majority of women, the picture was reversed. As an illustration he quotes a 'highly talented' woman asked which mattered to her more, family or work. She replied, 'What a question! The family is more important than anything. Look, my work is extremely important to me. I want very much to work, but I wouldn't invest one-fourth the thought to my work that I invest in my family, under no circumstances.'[11] Or a woman from another kibbutz study, 'Although the sexes may have the same abilities in art and science, men are usually superior because women are so absorbed in their basic responsibilities of child and home care . . . It is the most important thing in the world to her . . . It is a feeling of fulfillment. To give life is to overcome death.'[12]

I find the words of these *sabra* women immensely moving. They speak of choosing individual meaning in the face of a simplifying political ideology. To put it another way, they speak of Pan v. Spock.

Beyond family and work

The feminine revolution that Spiro documents in Kiryat Yedidim does not stop at family sleeping and work choices. To the horror, doubtless, of the idealist founders, in 1975 Spiro found that the kibbutz now had a 'beauty parlor, complete with beauticians, a skin specialist, and a masseuse'. Kibbutz women, dressed as field workers in 1951, now wore well-cut dresses, jewellery, stockings, high heels. As we shall see shortly, they were still committed equity feminists, but with an attitude like Chimamanda Ngozi Adichie in *We Should All Be Feminists*. She describes herself as 'a Happy African Feminist Who Does Not Hate Men And Who Likes To Wear Lip Gloss And High Heels For Herself And Not For Men'.[13]

By 1975, too, wedding ceremonies were back, along with a dramatic increase in the importance of the family residence. In a survey of 60 people across 6 kibbutzim, 'not even one understood why they were asked to comment on the reasons for solemnizing kibbutz marriages by means of a wedding ceremony'.[14] Attempts to avoid public displays of affection between husband and wife had been 'totally abandoned'. Now in the kibbutz, the expectation was back that young women needed a husband: 'to remain unmarried is a "tragedy"'.[15] Entering a kibbutz home, the visitor found that immediately 'the food is brought out – cookies, a cake, or a pie, and sometimes two kinds of cake or two pies – all of which the hostess has prepared herself'.[16]

The kibbutz pioneers had believed that family life breeds patriarchy. Committed to childcare, the woman is not free to take on leadership and power roles. Spiro's work shows that the pioneers – and the early socialists whose ideas they had adopted – were absolutely right. In the kibbutz, by 1975, the great majority of leadership positions – economic manager, general manager, member or chair of administrative committees – were held by men. But this was not because such positions were unavailable to women – quite the contrary. With occasional exceptions, the great majority of men interviewed were strongly in favour of bringing women into these roles. 'Not one among this majority', Spiro claims, 'entertained any doubts about women being intellectually qualified for these posts.'[17] By and large, women simply did not want them.

In part, this was just what the pioneers feared. Leadership roles in the kibbutz were sometimes full-time, or sometimes had to be undertaken after the workday. Once family sleeping was introduced, this simply did not fit with women's commitment to the family. Children now came home after school and their mothers cooked dinner; in the morning, they dressed them, cooked breakfast, took them to school. Childcare made it difficult for women to leave the kibbutz for advanced training or study. It was just what you might expect, and the women knew exactly what it meant – but just the same, they insisted. It is hard to resist the conclusion that, compared with women in other cultures (such as ours), they knew what most mattered to them exactly because they had been exposed to a world without it.

Family responsibilities were just one of the reasons *sabras* gave for avoiding these committee and leadership roles. Many told Spiro that positions like these made them uncomfortable, with 'a degree of stress or tension that they would prefer to avoid'.[18] Intriguingly, Spiro notes that, given the need for after-hours working and other burdens, along with at best modest power, many men were 'only slightly less reluctant than women' to take on these positions. But eventually they could be 'cajoled' – agreeing to serve, as Spiro puts it, 'only after consider-able (and often unpleasant) haggling'.[19] Later, I will come back to the argument that leadership is not a gift, but a life choice – one with its benefits, but also its costs.

Distributions

In Chapter 12, I showed a figure of distributions. For any char-acteristic we may measure, physical or psychological, the distributions of men and women may not be identical – but, of course, they heavily overlap. Spiro's *sabras* also understood this perfectly well. They do not say that all men value work over family, or all women value family over work. They speak of 'ten-dencies'. They recognise that each person is an individual, and that generalisations like these apply only to whole distributions. Here is one such. The woman has been asked, 'Do we have equal-ity of sexes in the kibbutz?' She replies:

> I think there are differences. There are physiological, bio-logical and psychological differences. I am not saying that women cannot be prime ministers or company directors. There are women, not many, who reach the highest pos-itions and fill them successfully. But as a rule women are not equipped to develop the way men do . . . It is the woman who bears children and is bound to them more.[20]

Going back to the figure of distributions, we might imagine that one shows the fulfilment felt from achievements at work, another the fulfilment from relations in the family. Spiro's *sabras* strongly believe that the distributions are not identical, but of course they are broad for both men and women. At one extreme, we all know many women who have pursued and loved brilliant careers. At another, we know men for whom family is all-important, and work a necessary nuisance. Again, these are the motivations and choices of individuals, not whole groups.

Spiro himself is quite clear on this. For example, he says, 'If kibbutz governance has become predominantly male, it is not because of lack of opportunity for, or encouragement of women, but because most of them – *there are, of course, many exceptions* – are not interested' (the italics are mine).[21]

In 1951, Spiro had conducted observations of preschool children playing in the unisex setting of their communal houses. Despite the way these children were being reared, he saw distinct, rather familiar, differences between boys and girls. Boys were much more likely than girls to be found playing with toys, especially large toys requiring strenuous effort; girls were more likely than boys to be playing fantasy or role-playing games, pretending to be somebody or something else. Among these role-playing games, boys were much more likely to pretend to be an animal – usually not one of the familiar domestic animals that they knew from the kibbutz, but something more exotic, such as a horse, snake or wolf. Girls were more likely to role play an adult female or a baby. But again, all of this was a matter of statistics. Intriguingly, for example, even boys were more likely to role play an adult female than an adult male – perhaps because the professional care-givers that they knew were largely women.

There is something else that was likely special in the kibbutz case. The weighting given to different innate releasing mechanisms may vary strongly between individuals, but also it depends strongly on triggering conditions. Undoubtedly, the child itself is the strongest trigger condition for the IRM of parental commitment and care, and in the kibbutz, very often, the mother could see the child but could not care for it. This is entirely different from a woman who decides never to have children because her career comes first. When we ask what fulfils us, the answer depends both on who we are and on the surrounding conditions of our own particular world.

Equal value

Spiro's interviews tell me that the kibbutz founders were wrong about gender feminism. It simply is not true that differences between men and women are culturally determined and can be eliminated by changing cultural practices. But quite inspiringly, the same interviews are a paean to equity feminism. Raised to be equally valued, men and women are more than ready to say yes, they are different, but yes, they are of equal value.

One woman explains that the difference between men and women is 'determined by nature . . . also true in animals'. She is asked whether it bothers her that long days of agricultural work are not consistent with the needs of the family. She says, 'Of course not. I think that being a woman has many advantages. Giving birth and all that is connected with it is something of which men never experience anything as profound. I have never felt that my sex was inferior, not ever.'[22] (Because I have known a few Israeli women over the years, I especially love that 'not ever'. When a *sabra* tells you 'not ever', you can be fairly sure she means it.)

Another:

In my view there are differences in the nature of men and women, and I see no point in denying them. But in the kibbutz this does not mean that men and women are not equal; just the opposite. For here both sexes hold the other in the same

THE ANIMAL AND THE THINKER

esteem . . . Of course, I am opposed to discrimination against women . . . If I believed it existed here, I would fight against it . . . But I don't feel it . . . On the contrary, I feel that I am the equal of any male.[23]

Spiro summarises it like this:

interviews with the sabras . . . indicate that they view males and females as completely equal in intelligence, in intellectual capacity, in their worth as human beings, and in their contributions to kibbutz society. They hold that sexual differences in social roles, including leadership roles, reflect differences in interest and needs rather than in talent or ability.[24]

I suspect that there are few cultures, including our own, that can say as much. In this sense, perhaps, the feminist aspirations of the kibbutz founders were not so unrealistic after all. They could not produce a revolution in human nature – but they could produce a revolution in human values.

Spiro began his work as a cultural determinist; many decades later, the lives and commentaries of his many *sabra* subjects had persuaded him that he was wrong. Let me give him the last word:

The kibbutz case does not prove the existence of precultural sex differences. Rather, it challenges the current intellectual and political pieties which deny the existence of such differences . . . on the grounds that to be different is ipso facto to be unequal. That individuals and groups must be identical in order to be equal is surely one of the more pernicious dogmas of our time, and the fact that, ironically enough, it has become a liberal dogma does not make it any the less so.[25]

The leaky pipeline

As regards work and political roles, the phenomena that Spiro documented in Kiryat Yedidim reflect burning issues in modern Western culture. Across a very wide range of occupations, men and women begin together on their careers, often in quite equal numbers, but through advancing levels of success and seniority, the proportion of women decreases.

As one example of this 'leaky pipeline', a study of the UK scientific workforce, published in 2014 by the Royal Society, showed that, averaged across areas of science, around 65 per cent of undergraduate students were women. There were the usual variations across discipline, with a minimum of 9.6 per cent in Mechanical Aero and Production Engineering, and a maximum of 90 per cent for Nursing and Paramedical Studies.[26] But by professor level, the average proportion for women was only 16.7 per cent, ranging from 5 per cent for Mechanical Aero and Production Engineering to 58.2 per cent for Nursing and Paramedical Studies.

Closely related to the disparity in numbers is disparity in salaries. In very many occupations – though as we shall see, not all – men and women carrying out the same job do not earn the same salary. In the Western world, such disparities have radically reduced over the past century but, very often, a substantial difference remains.[27]

On the one hand, doubtless such disparities can be manifestations of 'the patriarchy', fed by a wide range of gender-biased beliefs and practices. There might be a simple belief that women are not up to the job, sometimes taken to be conscious, sometimes (more insidiously) unconscious. There might be organisational practices that favour men's careers, such as preferring to hire men or more often putting men forward for promotion. There might be men's sense of unease when a woman is brought into the boardroom. Undoubtedly, such barriers to women are a part of the story of the leaky pipeline – sometimes, very likely, a large part.

If we think of Spiro's findings, however, it seems unlikely that this kind of barrier is the whole story. Surely, a part of the leaky-pipeline phenomenon must also be the motivational differences that were so vivid in the kibbutz. There is a question of work–life balance. Though our own work and family conditions are very different from those of the kibbutz, still it remains true that another hour spent at work is an hour spent away from the family . . . perhaps many women simply decide that perpetual struggle for career success is not what matters most to them.

Some time ago I attended a leaky-pipeline discussion on scientific careers at the Royal Society, chaired by a rather wonderful, extremely successful, extremely balanced-seeming woman. She asked the committee, 'Why do these women fail?' I said nothing, but my first thought was, *We don't know they 'failed' unless we ask them. How many actually liked where they went?* Science can be one good way to spend your life, but nobody could believe it is the only good way. In Western culture, correspondingly, the arrival of children is one of the commonest reasons for a woman to defer or abandon her previous career.

It would be incautious to generalise straight from the kibbutz case to our own case. The problem of work–life balance is a matter of adjudicating between competing calls, including the competing IRMs of family life and successful achievement in one's career. Another point that Spiro makes quite clear is that, in the kibbutz, many women – and for that matter, many men – were deeply dissatisfied by their work. In the kibbutz of 1975, work opportunities were extremely limited in terms of variety and challenge – a woman who had received a degree in chemistry might find herself working in the nursery, or some other role equally unsuited to her talents and training. In our own world, for many women and men, the interest and satisfaction of work will be far greater – very commonly, among the strongest interests and satisfactions that our world can offer. Again, the rival appeal of competing IRMs will depend on the circumstances, and our circumstances are very different from the circumstances of Kiryat Yedidim.

In the chaotic world of Pan and Spock, many factors will contribute to the differences of men's and women's careers. Bias and discrimination can be obvious, but even if these can be eliminated, it seems highly unlikely that men and women, considered as whole distributions, give the same weight to rival motivations and values – even outside the rather special world of the kibbutz. It's complicated, and the relative importance of different factors will surely vary from one case to another. And it matters, because a society that wants to address leaky pipelines needs to know why they happen.

Why motivations differ

Our motivations, I have argued, come heavily from Pan. In previous chapters, we have dealt with two of Viktor Frankl's core ideas on the sense of meaning. The first is the meaning we get from human relationships. The second is the meaning from struggle and achievement. The idea that we take from Spiro's *sabras* is that women and men, on the whole, give different weighting to these two great sets of IRMs and the sense of meaning they deliver. Women, on the whole, care more about their lives with other people – especially their own children – while men, on the whole, are more concerned with struggle and status.

Interestingly, Spiro's conclusions on this difference mirror very closely the conclusions that Eibl-Eibesfeldt drew from his studies of small-scale tribal societies. Of his *sabras*, Spiro says, 'The gratifications they derive from this "feminine" role [their role as mothers] obviate the need to strive for status in "masculine" roles.'[28] Eibl-Eibesfeldt says,

> the woman's role . . . is not saddled with insecurities. She proves herself to everyone by bearing a child. Thus the woman has a natural position of power, while the man has to establish his authority in confrontation with others, whether that be hunter, warrior, planter, or leader. He is selected for this kind of assertion and requires the continual recognition of others for his achievements. This characteristic is by no means absent in women, but they are less dependent on it than men, whose striving for recognition is much more pronounced.[29]

Even Margaret Mead, in her later writings, came to very much the same conclusions, speaking of the sense of 'irreversible achievement' that is brought by child bearing, while for men to be 'at peace, ever certain that their lives have been lived as they were meant to be, they must have, in addition to paternity, culturally elaborated forms of expression'.[30]

This is not at all to say that bearing and raising a child is the only way for a woman to be fulfilled, or that it is always the most important way. In our society, birth rates fall as more and more women devote more and more time to career. Again, there are many competing sources of fulfilment, with weights varying across people and time. But few people would question that, for very many women, bearing and raising children is a big one, often the biggest.

As we saw earlier in the book, Eibl-Eibesfeldt believed that, in the evolutionary development of individual pair bonding, the starting point was the bond between mother and offspring. In Chapter 4, I touched briefly on the neurobiology, worked out especially in rats and mice: the core role of the hypothalamus and amygdala, with their central role in so much instinctive behaviour; the massively complex picture of hormonal and neural events, with responses to many inputs from the rat or mouse pups – their smell, their suckling – inhibiting the mother's avoidance and promoting the specific behaviours needed to care for the young.

In women, similarly, there are enormous and complex hormonal changes through pregnancy and after birth, and though the resulting changes in the brain are less well understood than they are in rodents, in our species too such changes promote the mother–child bond.[31] We know that a rush of love can accompany breastfeeding, very likely associated with skin-to-skin contact and the release of hormones including prolactin and oxytocin. After birth, there is increased grey matter in several regions of the mother's brain, including the hypothalamus and amygdala, and these increases can be linked to the mother's positive feelings about her baby. Oxytocin levels predict sensitive mothering and affectionate contact. When the mother is exposed to cues from her infant, brain imaging shows increased hypothalamic and amygdala activity. These events concern the bond between a particular mother and her infant, but evidently, women's strong interest in babies and children is far more general than this. In my workplace, and I imagine in most others, a woman who brings her baby to work is immediately surrounded by a group of (mainly) other women, admiring and wanting to play. I hate to admit it, but the reason I don't stop and join the group is that, if I am passing in the corridor, I am already doing something – and perhaps sadly, the attraction of the baby is simply not sufficient to distract me from whatever that something is. A very large research literature confirms the conventional wisdom that interest in other people, in caring professions, in the concerns of their children, is stronger in women, while interest in gadgets, outdoor work and so on is stronger in men.[32]

Also rather familiar is the idea that, on the whole, males are more driven by the needs for struggle, display and recognition for some form of excellence (from sweet potato bigman to nerdy computer hacker). In any primary school classroom, it is obvious which children are showing off to their peers, holding court and disrupting the proceedings. Discussions of how to better involve women in group discussions emphasise how men posture and shout them down (for that matter, very often they are posturing and shouting down everybody).[33] We call it 'willy waving', and it is rather reminiscent of the vervet monkeys from Chapter 8, standing guard around the troop and displaying their vivid red erections to ward off an aggressor.

In Chapter 11, I suggested that Frankl's sense of meaning through achievement is built heavily on the IRMs of aggression. In part, this is overt competition with others, for recognition and a place in the pecking order, but in part too, I suggested, it is exaptation that makes an enemy of the challenge itself, the problem to be attacked. If this idea has merit, it would be rather unsurprising that struggle means more to the more aggressive male, built for competition and display. A female colleague and dear friend once told me how impressed she was by her male counterparts who seem so certain how great they are. I told her that this certainty is no more than a peacock expanding its enormous tail – and that you should never actually believe the peacock but, on the other hand, you would scarcely want him to lose his tail.

Modern discussions of the workplace will often decry 'toxic

masculinity', and a competitive, laddish work culture than many women find unappealing. Not surprisingly, research shows how strongly we need a sense of belonging at work – a sense that we fit, that we are valued, that the team knows our importance.[34] Equally, research shows the cost of alienation, and the sense that the work environment is not suited to the person we are. Obviously, a heavily masculine workplace will not appeal to people who are not heavily masculine themselves. Very likely, discouraging the willy waving can make workplaces more inclusive and, undoubtedly, more inviting for many women who do not especially enjoy the struggle. It is often thought that a diverse workplace has advantages not only for the diverse individuals it welcomes, but for the diversity of ideas brought to the work itself.[35]

As always in the world of Pan and Spock, however, complex reality has opposite sides. The desire to display and struggle – especially strong in men – may often need to be reined in; but at the same time, I argued earlier, it is an engine forcing human activities forward. There is a reason our aggressions were exapted for this purpose – humanity needs achievements and, in my opinion, the love of battle is a major force driving us to attain them. It is not just that people – especially men – do it. It is that they *love* it, and because they love it, they do it all the harder.

Teenage boys are well known for their over involvement in road crashes, and this is a phase I remember well. For the first few years after I learned to drive, I hurtled the battered farm Land Rover (partly battered because my older brother had already learned to drive in it) as fast as I could round the winding Devonshire country roads, trying to find out exactly how fast I could take each one of the many familiar bends. Who was I showing off to? I was on my own. Who was the enemy? The enemy was simply the universe, and the joy was the joy of showing the universe what I could do.

Accident rates suggest that this kind of thinking is heavily masculine, and I can't help suspecting that, if my reader is a woman, she will be left thinking (correctly) that young men are morons who should not be allowed on the roads, but if he is a man, he will be unable to resist a slightly shame-faced smile of nostalgic recognition.

Once, in my office at work I made a discovery. Never mind what it was, or how I made it – as I plotted the data on to a sheet of paper, I realised it was big. The more I plotted, the bigger it got. Eventually I snapped, grabbed up the paper and ran out into the corridor to find someone – anyone – to show. The first person turned out to be my co-conspirator on the project; I waved the result under his nose and yelled, 'We're going to be so fucking famous!'

I promise you, if you interviewed my students, collaborators and colleagues, they would tell you I am one of the gentlest and most committed scientists they know, a real old schooler whose only goal is to do the best job he can. But inside, Pan is yelling. I am doing the best job I can, not for the abstract sake of knowledge, but because it just feels so good. Like that dog on the beach at the start of Chapter 11, we strive because we were built to . . . we get the joy, and humanity gets the results.

And once again, any difference between men and women is all just statistics. Of course, I know heart-stoppingly brilliant and successful female scientists, at least as determined as I am to rip into their problem and show the universe what they can do, and quite often better at it. It would be absurd to think that the average man has anything like the competitive drive of a professional female athlete or sportsperson, with sparkling achievements that few men could ever hope to emulate,[36] or, for that matter, the competitive drive of my own mother-in-law, who would never let the grandchildren win at cards even when they were in kindergarten. But if we are looking at proportions of board chairs or university professors that are female, then the question is statistical, and it is the whole distributions that matter.

Any woman who has wondered why men can be so childish is going to enjoy this, again from Eibl-Eibesfeldt:

> Chimpanzees use rocks to break open nuts, and this is accomplished particularly by the females, who are more skilled at the task. Their strokes are more even, indicating superior fine motor control . . . Males often strike too powerfully, smashing both shell and contents . . . In the Japanese macaques, females were the inventors of new traditions of food processing, the males being too preoccupied with territorial defense and rank displays.[37]

Preoccupied with territorial defence and rank displays . . . were they now?!

How salaries can become equal

In 2023, the American labour economist Claudia Goldin won the Nobel Prize 'for having advanced our understanding of women's labour market outcomes'. For generations, there has been convergence between the salaries of American women and men carrying out the same work. There has been convergence in career choices, education, qualifications obtained. Despite all this, in many lines of work, significant salary discrepancies remain. Goldin asks why and, most significantly, she asks what would be needed for these remaining discrepancies to be eliminated.[38] Her conclusions are perhaps surprising. The solution, she notes, need not involve government intervention.

It need not make men more responsible for work in the home (although, as she wryly comments, 'that wouldn't hurt'). Instead, salary discrepancies depend on a particular structure of the work itself – and if they were finally to be eliminated, it would be this structure that would have to change.

Goldin's analysis is based on the comparison of different occupations. In some occupations, such as business and law, salary increases exponentially with hours put in. The person who works twice the hours is paid much more than twice the salary, and for the person who puts in four times the hours, the increase is correspondingly larger again. For other occupations – Goldin uses pharmacy as an example – it is much more linear; the person who works twice as long gets twice the money. Using a comparison of demands in different occupations, Goldin concludes that the key variable is whether one person's labour can freely be exchanged for another's. When this is true, the relationship between hours and work is roughly linear, but when it is not, massive rewards attach to working longer hours.

Why is this? If a lawyer has spent months or years forging a relationship with a valued client, understanding the complexities of their business and their legal needs, then it is this particular lawyer who needs to put in the hours on this client's business. They cannot go home early and pass on the task to somebody else. But in careers such as pharmacy, computerised records tracking patients' needs across different branches, hospitals or insurance companies mean that pretty much any competent pharmacist is as good as any other. This was not always the case; when pharmacies were small and independent, each pharmacist needed to know the individual client and their needs, but as chains have grown and records have become shared, this need has evaporated. In parallel, the salaries of men and women have converged, putting this 'among the most egalitarian' of high-income American professions.

On this analysis, once again, the key consideration is work-life balance. Women want or need more flexibility in their working lives, very commonly because of their children, and as long as this remains true, their value in the 'exponential' professions is inevitably reduced. A person working part-time is less certain to be available for those aspects of the work that, critically, only they can carry out effectively. The idea that an employee has individual values goes deeper than this, too. It also matters how many hours they have put in over their whole working life, with the specific value that these hours have given them. This will always be important if a part of the person's value is the specific knowledge that they, and only they, have acquired through their career experience.

If women often need or want more flexibility in working hours, then when value and reward increase exponentially with time put in, gender discrepancies in salary will remain. (Again, of course, this is a matter of statistics, not individuals. On this analysis, the career woman who puts in just the same lifetime hours as a matched career man should not suffer at all. If she does, this may well be the insidious work of the patriarchy.) Goldin's conclusion is that discrepancies will only go away if jobs themselves can be changed, eliminating the role of individual value. For pharmacy, this has proved possible, women very frequently work part-time, and gender discrepancies in salary have all but disappeared. For many other occupations, though, it is hard to see how it could happen. As Goldin puts it, 'There will always be 24/7 positions with on-call, all-the-time employees and managers, including many CEOs, trial lawyers, merger-and-acquisition bankers, surgeons, and the US Secretary of State. But, that said, the list of positions that can be changed is considerable.'[39]

Barriers

None of this, of course, means that equity feminists are wrong when they point to gender bias, men's defence of the patriarchy and the belief that women 'don't belong' in senior roles. Needless to say, this happens, often a lot. As women join the workforce, it is unsurprising that the men already in that workforce can feel threatened, or that things have changed for the worse, or that the power and privilege they are accustomed to will be taken away. We have expressions such as 'wearing the trousers'. We have the 'fragile male ego' and the competitive male's fear of being beaten – especially by a member of 'the weaker sex' (a fear, very likely, based on a long evolutionary history of physical conflicts and a physically weaker sex). We have the failure to elect Hillary Clinton, which surely, in my opinion, was influenced by reluctance to support a woman who was just *too* good, *too* qualified, *too* obviously the right candidate for the job. In many cultures, including our own until the suffragettes, we have explicit mechanisms of discrimination, fully prohibiting women from roles of power and influence.

While this discrimination is patently unfair, violating Fiske's equality matching, it is easy to see how, in the male mind, influences from Pan lend their support. In the authority ranking mode, nobody wants to lose their place in the pecking order. Nobody wants to see their privileges taken away and given to somebody else. We have the strong human tendency to separate into 'us' and 'them', promoting the agenda of 'us' and believing that 'they' are lesser, even worthless.

Sometimes this discrimination is comically absurd. In March

2021, the Suez Canal was blocked by a grounded container ship, the *Ever Given*. The BBC ran a news story on Marwa Elselehdar, the first Egyptian woman ever to become a ship's captain.[40] On social media, a rumour began to circulate that Marwa Elselehdar was the captain responsible, though in fact, she had nothing whatsoever to do with it. She was hundreds of miles away, working in Alexandria. Behind this absurdity, however, was the story of a woman smashing through ingrained barriers of expectation and belief. She explained that she had 'always loved the sea, and was inspired to join the merchant navy after her brother enrolled . . . Though the academy only accepted men at the time, she applied anyway and was granted permission to join after a legal review by Egypt's then-president Hosni Mubarak'. In her training she was surrounded by 'older men with different mentalities'. 'People in our society still don't accept the idea of girls working in the sea away from their families for a long time,' she added. 'But when you do what you love, it is not necessary for you to seek the approval of everyone.' This is a real life story of *Jane and the Dragon*. In 2017, Marwa Elselehdar was honoured by Egypt's then-president as part of Egypt's Women's Day celebrations.

Often, of course, the story of discrimination is much darker than this. In the United States, Susan B. Anthony fought her whole life to win the vote for women – and died in 1906 before her aspiration was finally realised. In the news, we read constant stories of how women in the London's Metropolitan Police, or the fire service, or professional football, leave their careers because their progress is continually blocked and their ability belittled. All of this is palpably, often bitterly unfair, and very much under cultural influence.

As shown by the history of our own culture since the nineteenth century, and by the continuing changes today, discriminatory beliefs and practices can be very substantially changed by protest, education and discourse. Like any other application of Spock to adjust human life to its changing conditions, this does work. On its own, though, it may not be enough to eliminate leaky pipelines – precisely because this is the Pan of different natures as well as the Spock of different expectations and beliefs. It makes sense to attack a culture of patriarchy and to struggle to change it. These struggles, however, will surely be more wise – and quite possibly, more effective – if we recognise that the cultural patriarchy does not exist in a biological vacuum.

As I already mentioned, there is a strong leaky pipeline in academia, with only small proportions of women reaching professorial positions. I have sat in many meetings discussing what should be done, and the central idea is always bias. In Chapter 12, I mentioned the charming TV series *Endeavour*, with its young detective learning his trade in 1960s Oxford. Periodically in this series, we encounter some pompous Oxford don explaining that women are not suitable for academic life, and to me it seems perhaps fair but at the same time exaggerated. Since my arrival as an undergraduate in 1970, I have spent my life among Oxford and Cambridge academics, some of them quite pompous, but I think I can put my hand on my heart and say that nobody, ever, has expressed (in my hearing) the belief that women are not suited to the job. Of course, it happens – friends tell me that they have certainly encountered it, and perhaps it varies between disciplines – but it is sufficiently rare that in over 50 years, it has never happened around me. It would actually be a rather stupid belief to hold, since all around us are brilliant women doing the job brilliantly.

A familiar further step is to invoke 'unconscious' bias, and I am sure there is something in this too. In a typical experiment, reported by a team of communication scientists led by Silvia

Knobloch-Westerwick, graduate students were asked to rate the quality of research abstracts prepared for presentation at a scientific conference.[41] The same abstracts were given (fake) male or female authors, and when the author was a male, the rated quality was higher. The difference was reliable – but tiny, less than 1 per cent of the whole rating scale. Perhaps many small influences like this add up over a woman's career, putting the brakes on her success. Many organisations – including the University of Cambridge – now require 'unconscious bias training', alerting employees to the possibility of such biases. Such training is cheap to implement, and perhaps gives the organisation a sense that it has ticked the box of addressing its leaky pipeline. Trials, however, suggest that unconscious bias training in fact achieves very little – either it is just not very good, or (as I suspect) it is often addressing the wrong problem.[42]

Certainly there are much bigger effects to be found, and not surprisingly, they remind us strongly of the competitive, willy-waving male. This time, biased beliefs are beliefs about ourselves. In an enormous recent survey, the social scientists Clotilde Napp and Thomas Breda asked half a million 15-year-olds from 72 countries to describe themselves.[43] Across countries, 61 per cent of girls – but only 47 per cent of boys – 'agreed' or 'strongly agreed' with the statement, 'When I am failing, I am afraid that I might not have enough talent.' The difference was seen in 71 of 72 countries; intriguingly, it was *strongest* in countries like Iceland and Denmark, with the greatest objective gender equality in labour market, education, health and political representation, and weakest in countries such as Saudi Arabia and Lebanon, with objectively strong gender discrimination. The huge sample allowed this difference in self-belief to be related to students' actual ability, as measured by school achievement; it was stronger among *more* able students.

All of this suggests that, when boys and girls are put into direct competition, and especially when they are doing well, girls judge themselves less confidently. Boys indeed were much more likely to agree with 'My belief in myself gets me through hard times,' and reminding us again of Spiro's *sabras* (and of much other research), 'I enjoy working in situations involving competition with others.'

It is often suggested that the leak in the pipeline may be lessened by the inspiration of successful, powerful female role models, and indeed, this is what Napp and Breda conclude from their findings on self-confidence and self-belief. A few years ago, I read the obituary of an old friend, Leslie Ungerleider, an esteemed and enormously influential neuroscientist. The (female) author said, 'She had unwavering trust in the talents of her female mentees and instilled confidence, telling them "You can do it!"', and reading this I could just hear Leslie's indomitable voice.[44]

In 2021, when Kamala Harris became the first Black and Asian American woman to be made vice-president, the BBC ran a story based around a picture of her and two female friends, taken many years previously at her alma mater Howard University.[45] Howard was historically a Black school, founded to educate Black Americans after the abolition of slavery. One of the friends in the photograph was Valarie Pippen, who went on to medical school. Of Howard, she said, 'We all had . . . a striving to do well, a striving to live with integrity and to make your mark on the world.' The other was Karen Gibbs, now an attorney. Of her professors, she says, 'They were realistic to tell us what we would confront when we left Howard – but they equipped us to realise and achieve our dreams.' These students were potential underdogs because they were both female and Black, but their lives tell us that inspiration really can work.

Realising our dreams

Karen Gibbs speaks of achieving our dreams, and equity feminism is all about dreams – a cultural imperative that everybody gets the same chance for their dreams to be realised. Gender feminism thinks that the dreams themselves are a cultural invention, and that if men and women are brought up the same, their dreams will also be the same. Dreams, however, come heavily from the world of Pan, with the beckoning IRMs that demand to be discharged. Everybody has their own, but their distributions are not the same in men and women, and neither should we expect the resulting lives to be the same.

As Christina Hoff Sommers points out, gender feminists often have a strong taste for social control. She quotes Simone de Beauvoir, once asked whether women who wished to should be permitted to stay home raising children. 'No,' she responded, '... women should not have that choice, precisely because if there is such a choice, too many women will make that one.'[46] Such messages carry all the usual risks of Spock, as a political ideal imposes its own over-simplified image on complex reality – forcing individual women to struggle with a life that, for them, is just not the most important one.

There can be a lot of wisdom in children's fiction, and to *Jane and the Dragon* let me add Enid Blyton's series *The Famous Five*. The Famous Five are two boys, Julian and Dick, the dog Timmy, and the two girls George and Anne. It is the character of George that really makes these books special – brave, short-tempered, always ready to prove she is 'as good as a boy', always first into whatever hidden underground tunnel the adventure discloses (in *The Famous Five*, there always seems to be an underground tunnel). Make no mistake, Anne has the spirit of adventure too, and on her part too there is not the slightest reluctance to go down into a tunnel – but by and large, she is at least as happy making the sandwiches for the group's picnic and handing them around when the hamper is opened. A book with only two Annes would be anodyne, but a book with only two Georges would be unreal. Both characters seem like real people with real dreams. As Spiro's 'Ruth' says, it is perfectly possible for the life choices that these two girls will make, as well as the life choices that await the two boys, to all be considered 'equal' – just not the same.

PART V

Power

CHAPTER 15

There is no will of the people

Surely this should be Spock

In our complex modern world, we might hope Spock to be at the core of political decisions. When decisions affect communities of millions of people, whatever else they are complicated, and far beyond the hunter-gatherer world that shaped the promptings of Pan. Surely we need our enormous bank of real-world knowledge to decide whether Britain should remain in the European Union, or what arms should be supplied to a struggling Ukraine. Even with this knowledge, decisions are wildly uncertain, with the situation just too complex for accurate predictions. If we do not use everything we know, it seems obvious we have no chance at all.

Ideally, perhaps, such decisions would be handed over to the perfect Spock of *Star Trek*, with all knowledge at his fingertips, and the ability to integrate all relevant information in the service of the best possible decision. At least for now, this ideal Spock does not exist, but surely our best bet is to lean heavily on the human Spock we have, with all his ability to wrestle with the novel problems of today's world.

At first sight, democracy seems not a bad way to achieve this.

Many millions of individuals have their thoughts and, by a system of voting, these thoughts are combined. It sounds good . . . but in democracy, all too often, the combination of thoughts does not work, and even Spock's best thoughts are overruled by Pan.

There is no will of the people

As we have repeatedly seen, in the world of Pan and Spock, a person is not a single, coherent whole. A person is a mass of conflicting thoughts and needs, with elaborate, often conflicting innate releasing mechanisms demanding to be satisfied, and innumerable tangled ideas, often quite in conflict with one another. The person is a patchwork, with no fixed things that they 'think' or 'want'. What they think or want changes from moment to moment, depending on which IRMs are most activated, and which ideas have currently been brought to the front of the mind.

It is tempting to think that, even though the ideas of an individual can be moved around by time and circumstance, the ideas of the people will still remain good. At the heart of democracy is 'the will of the people'. By casting his or her vote, the individual indicates their will over what is the best choice – either the best choice of government or, in the case of referenda, the best choice on some specific social issue. By combining the votes of millions of individuals, we can approach the real truth of what the body as a whole 'wants'. In his inaugural presidential address, the Venezuelan populist Hugo Chávez put it like this: 'All individuals are subject to error and seduction, but not the people, which possesses to an eminent degree of consciousness of its own good.'[1]

This idea is tempting, but it is built on a statistical fallacy. In Chapter 5, I discussed the evolution of ideas and, in particular, the instability that results from mutual influence. If voters independently formed their views of 'their own good', then indeed, with enough voters, we might approach a truth of what was really best for the majority. This would apply as long as, on average, beliefs were more often right than wrong.

But as I explained in Chapter 5, this all breaks down once the opinion of one person influences the opinion of others. Now the whole system is highly unstable – as soon as one person forms an opinion, others tend to follow suit, and the whole process runs away with itself. This happens with any culturally transmitted belief, and evidently, it happens a great deal in the case of political decision-making and democracy. We now have millions of people forming an opinion on some issue such as the merits of Brexit or Donald Trump. Leaders compete to bring people to their side. Each person who swings one way encourages others to follow. Democracy is not a system for balanced decision-making. It is a system for massive and irrational instability. As Winston Churchill said, this may still be better than any alternative. But especially with the forces of Pan at work, and especially with our ever-increasing ability to influence one another, not just our immediate acquaintances but potentially millions of others, this is a system of decision-making that can run rapidly out of control.

It seems obvious, too, that this unstable vying for power must be exaggerated in modern democracies, with their organisation into competing parties. This modern system is quite different from the original, Athenian democracy, reaching its peak under Pericles in the Golden Age of Athens in the fifth century BC. In Athenian democracy, there were no parties, though of course there would be blocs of opinion. There were no elected representatives competing for the right to make political decisions. Policy was formed by majority vote, with perhaps 30,000 male citizens eligible, and a quorum of at least 6,000 required for debates on a major policy decision. At these debates, any citizen was free to speak. This may not be practical in today's much larger societies, but it is a democracy very different from ours, with different demands on human decision-making. One introduction of Athenian democracy suggests,

> Pericles would have been . . . scathing of Western-style representative democracies that he would undoubtedly have seen as anti-democratic. The key to democracy, as far as 5th century BC Athenians were concerned, was active participation by the citizen body in all political aspects of the polis.[2]

No system of joint decision-making can be entirely free of the instability built in when the words of one person can influence the thoughts of many others, but surely a party system – especially when there are only two competing parties – must be about the most unstable there could be.

All systems with rival leaders and political parties must have the same problem of instability, and a tendency for 'the will of the people' to run away with itself. This seems especially dangerous, however, when the ideas of Spock are discredited, leaving the field open for Pan. Doubt over 'expertise' and 'knowledge' are a defining feature of populism. In their book *Populism* – which I shall use a lot in this chapter – Cas Mudde and Cristóbal Rovira Kaltwasser define populism as an ideology 'that considers society to be ultimately separated into two homogeneous and antagonistic camps, "the pure people" versus "the corrupt elite", and which argues that politics should be an expression of the [general will] of the people'.[3]

In modern political discussion, we hear a great deal about Washington and Westminster elites, who are out of touch with real people, and who cannot be trusted to have anything but their own interests at heart. In the Brexit wars, we heard constant arguments against so-called experts. Right-wing populism in the United States takes aim at a cultural elite, claiming they use education to pervert the views of lawmakers, politicians and public commentators.

Mudde and Rovira Kaltwasser suggest that the populist leader will project himself (it is usually a man) as a man 'not afraid to take difficult and quick decisions, even against "expert" advice'. He draws on 'anti-intellectualism and a sense of urgency'.[4] In Peru, the populist Alberto Fujimoro was elected president in 1990. He presented himself as a 'pure' person, who wanted to get rid of the corrupt elite. One of the slogans of his campaign was 'A President Like You', a subtle attack on his main opponent, the famous writer Mario Vargas Llosa, who won the Nobel Prize in Literature in 2010.[5] In his 1987 novel *Anthills of the Savannah*, Chinua Achebe draws an ominous picture of an African leader who discredits education and intellectualism precisely because it is the educated and intellectual who will oppose the oppression of his regime.

The use of the word 'elite' seems clever to me. Elite can mean privileged, as in the degenerated, childish surface dwellers of the future in H.G. Wells's 1895 novella, *The Time Machine*, barely aware that their civilisation depends on the turbulent, dangerous workers who live out their lives in the tunnels below. But elite can also just mean 'the best', as in elite troops. When it comes to navigating a path through the complex maze of an economic or political decision, surely knowledge must help and the informed expert is elite in the second sense. Beginning with theorists such as Max Weber and Joseph Schumpeter in the late nineteenth and early twentieth centuries, a strong movement in political science has been 'elite theory', maintaining that powerful, informed elites are both inevitable and necessary in a practical political system.[6] But in our world, educated experts will usually be privileged too, with more money, more interesting jobs and probably better health and longer lives. It is easy to raise resentment against a person who is more privileged than you, and for the populist leader promoting a questionable agenda, the word 'elite' can nicely transfer this resentment to the experts who might do the questioning.

Nobody could seriously believe that the world of Spock – the world of knowledge and expertise – is in general a bad thing. When I hear somebody question expertise in general, I want to say, 'Really? So next time your car breaks down, you'll take it to the dentist, and next time you have a toothache, you'll go to the garage?' Of course, we need Spock's elites! – in every corner of our lives.

In politics, though, it is all too easy to raise distrust of people who 'think they know more than you do' – they are often privileged, they may not have your best interests at heart and, quite often, their expertise will fail and they will get it wrong anyway. But still, it is a brave person who, deliberating on a political or economic question, and recognising their own ignorance of its complexities, still believes that it is their own opinion they should trust.

Pan writ large: The enemy

Leaders do not just formulate policy. In a democracy they need also to win support. As parties and leaders compete for power, it is Pan that we see writ large.

As we saw in Chapter 8, by analogy with a flock of greylag geese uniting their gaggling heads against a common enemy, Lorenz saw in humanity a pattern of 'militant enthusiasm' – a passionate coming together against some threatening 'other'. The deep sense of unity, of shared struggle for survival, can unite millions of individuals. Politicians know well, of course, how to use this aspect of Pan. At his trial in Nuremberg, Hermann Göring said:

> Naturally, the common people don't want war: Neither in Russia, nor in England, nor for that matter in Germany. That is understood. But, after all, it is the leaders of the country who determine the policy and it is always a simple matter to draw the people along, whether it is a democracy, or a fascist dictatorship, or a parliament, or a communist dictatorship. Voice or no voice, the people can always be brought to the bidding of the leaders. That is easy. All you have to do is tell them they are being attacked, and denounce the peacemakers for lack of patriotism and exposing the country to danger. It works the same in any country.[7]

The enemies can be from outside, like the Mexican hordes who must be kept out by a splendid wall, defending against 'animals', 'drug dealers', 'rapists'. They can be political opponents, like the leavers and remainers of the Brexit referendum. They can be Jews, gypsies or the welfare cheats castigated with such satisfaction by parts of the British press. In present-day Europe, a common move by radical right-wing parties is to manufacture an 'immigration crisis', which surely never touches the lives of the vast majority of voters. Mudde and Rovira Kaltwasser give the example of the True Finns who, after the global financial collapse of 2007, obtained an 'astonishing' 19 per cent of the vote in the elections of 2011. They claimed that their 'generous welfare state' was threatened by 'an "invasion" of immigrants'.[8] I am very fond of an American greeting card that circulated on social media during Trump's presidency, reading, 'My ancestors didn't come to this country to have it overrun by immigrants.'

In a battle of us versus them, we want to be the winners. I

vividly remember, in 1982, how, during Britain's conflict with Argentina over the Falkland Islands/Malvinas, every day the BBC news carried a count of how many British and Argentine planes had been shot down, and I really did feel that I was watching goals scored for and against in a football match. This conflict saved the political career of Margaret Thatcher, whose party was all but dead in the polls before she sank the *Belgrano*. Trump says he will make America great again, his supporters will be winners, they'll win so much that they get tired of winning. It is no more subtle than a phalanx of geese craning their necks together and gabbling – but it does work, it works beautifully.

Intriguingly, Eibl-Eibesfeldt speculates that the enemy does not have to be human – it can be chaos, with the strong leader offering a sense of order.[9] Benito Mussolini's Fascist propaganda machine boasted of trains running on time. In the Brexit referendum, the probability of choosing 'leave' increased systematically with voters' age. It was sometimes said that older voters wished to return to the time of a better Britain, but I was there at that earlier time, and it wasn't better at all. When I grew up, Britain was still emerging from food rationing – we had National Health Service orange juice and cod liver oil to protect children from vitamin deficiency, we had hand-me-down clothes and no TV. It is complete speculation on my part, but given these Brexit data, I couldn't help wondering whether, for older people, the vanished world is just a comfort. In our twenties, we learned the way of the world and gained mastery over it; in our seventies, we buy air tickets on websites and must decide which emails are phishing attacks. Of course, there is a sense that control has gone, with globalisation of almost everything a major culprit, and perhaps older voters were just objecting to this new, challenging, often foreign world.

The global power of 'find the enemy' is shown by the lifetime work of the right-wing pollster, consultant and strategist Arthur Finkelstein.[10] Finkelstein was a rather secretive figure, credited with the electoral success of many Republican politicians over the decades from his early work in the 1960s until his death in 2017. Finkelstein worked early on with Richard Nixon and Barry Goldwater, and over the years was credited with the success of Ronald Reagan, the union of Republican and Christian values, and other broad themes of modern American politics. As his primary threatening enemy he used 'liberals', and is taken as the person who managed to make 'liberal' a dirty word (an achievement indeed, as 'liberal' must certainly be one of the more positive words of the English language). He pilloried rivals as 'ultraliberal', 'superliberal', 'embarrassingly liberal', 'foolishly liberal', 'unbelievably liberal'. His straplines were things like 'Bob Abrams: Hopelessly liberal' or 'Mario Cuomo: Too liberal for too long'. His victims called it Finkel-think; his enemies called him 'The Merchant of Venom'.

In later years, Finkelstein helped design election campaigns in many countries – Israel, Hungary, Albania, Austria and more. In Israel, he is credited with the campaign that brought Benjamin Netanyahu to power, with the slogan 'Netanyahu is good for the Jews', and scaremongering over the threat of the Arabs. Perhaps most remarkable was his use of the billionaire and philanthropist George Soros as the enemy needed to keep the right-wing Viktor Orbán in power in Hungary. With Finkelstein's help, Orbán had already been elected by a landslide, but the opposition was now so weak that it was felt a bigger, better enemy was needed. The Hungarian people were struggling from the after effects of the global financial crisis, and Soros was identified as the perfect believable enemy – a billionaire and speculator, internationally connected, Jewish. Articles began to appear describing a fictional Soros conspiracy against Hungary. Orbán spoke of the hidden plan to 'weaken the national state' and bring 'Western' ideas to Hungary. Now, around ten years later, Orbán is still in power, and on conspiracy websites around the world, the liberal-minded, philanthropic Soros is vilified as 'the antichrist' and 'the most dangerous person in the world'.[11]

Finkelstein did not make many speeches, but in one of his last he said, 'I wanted to change the world. I did this. I made it worse.'

Bringing people together

Finkelstein was a master of forcing people apart. To oppose this technique, political leaders may resort to promoting 'the bigger picture' – the sense that, different though we may be, we are all in this together. In the 2020 Donald Trump/Joe Biden presidential election, on behalf of Biden and against the arch-separator Trump, the actor Sam Elliott narrated a well-known 'one America' video. Against the background of scenes from the great American continent, Elliott says, 'There is only one America . . . no Democratic rivers . . . no Republican mountains . . . there is so much we can do if we choose to take on problems, and not one another.'[12] I have no idea whether this approach works, but the underlying idea is clear enough, strongly reminiscent of material in earlier chapters, and consciously avoided by the political divider. In *The Righteous Mind*, published in 2012, Jonathan Haidt describes how, in 1995:

> Newt Gingrich, the new Speaker of the House of Representatives, encouraged . . . incoming Republican congressmen to leave their families in their home districts . . . Before 1995, congressmen from both parties attended many of the same social events on weekends; their spouses became friends;

their children played on the same sports teams. But nowadays most congressmen fly to Washington on Monday night, huddle with their teammates and do battle for three days, and then fly home on Thursday night. Cross-party friendships are disappearing . . . scorched earth policies are increasing.[13]

As a technique for persuasion, what almost certainly will not work, tempting though it may be, is to pour scorn on the opposition. As Mudde and Rovira Kaltwasser put it, 'In many cases establishment actors launch a coordinated frontal attack on the populists. By collectively portraying "them" as "evil" and "foolish," the establishment actors play into the hands of the populists.'[14] In an us–them situation, nobody takes kindly to insulting epithets from the opposite camp. Following the Trump/Clinton election, there were memes purporting to show that the preference of each American state was predicted by the mean IQ of its population (no prizes for guessing in which direction). In the Brexit referendum, working-class voters were more likely to vote leave, and it was common to hear (non-working class) friends inveighing against the 'stupidity' of these people. This is not a great technique for persuasion and, arguably, not a great technique for understanding. It is Pan who so effortlessly allows people to sort themselves into us and them, and Spock with his advanced reasoning powers may have little to do with it. We end up in group A or B not largely because we reason ourselves there, but because of who our friends are and which newspaper we read.

Pan writ large: The leader

Along with a hated enemy, Lorenz suggested that the full release of militant enthusiasm requires the figure of an inspiring leader.

As he puts it, 'Even the most emphatically anti-fascistic ideologies apparently cannot do without it, as the giant pictures of leaders displayed by all kinds of political parties prove clearly enough.'[15]

As we discussed in Chapter 2, human IRMs, both for demanding a position as leader and for accepting and submitting to that demand, have the usual strong resemblances with the IRMs of chimpanzees and other primates. The person submitting may throw himself to the ground, or at least give a deep bow; in return, there is a gesture of approval and benediction, perhaps laying the hands on the recipient's head. The primatologist Jane Goodall famously remarked how strongly Trump's posturing reminded her of dominance displays among male chimpanzees – the stamping, the blustering, the swagger, the noise, the impression of large size exaggerated by upright posture.[16] Surely, as we saw in Chapter 7, the IRMs of human dominance derive in part from the parent–child relationship and, very often, leaders are called 'father of the people' and refer to their followers as their children.

Mudde and Rovira Kaltwasser are at pains to point out that the aggressive, virile, charismatic strongman is not the only kind of populist leader – but he is a common kind. The military background of South American populists, such as Argentina's Juan Perón and Venezuela's Chávez, projects iron will and potential violence. Faced with potential scandal over his sex parties, the Italian premier Silvio Berlusconi used the media attention to emphasise his virility, only strongly denying the accusation that he had *paid* for sex with call girls at the parties. 'For those who love to conquer, the joy and the most beautiful satisfaction is in the conquest. If you have to pay, what joy is there?' he once said in an interview.[17] Opposed by Hillary Clinton, Trump once tweeted, 'If Hillary Clinton can't satisfy her husband what makes her think she can satisfy America?'[18] Umberto Bossi, leader of Italy's right-wing Northern League, would tell crowds 'the League has a hard-on'.[19] During his election campaign, the anti-establishment president of Argentina Javier Milei showed he was the man to slash bureaucracy by waving a chainsaw at crowds.[20] His slogan was 'Long live freedom, damn it!'

Or if the leader is a woman, she may use the Pan of mother-hood. Pauline Hanson, founder of Australia's One Nation Party, said, 'I care so passionately about this country, it's like I'm its mother, Australia is my home and the Australian people are my children.'[21]

One essential move is to project the image of 'being on the side' of the supporters, or preferably the people at large, and again, this can be spectacularly successful, even when it is mani-festly absurd. Many populists, claiming to represent 'the common person', are among the very wealthy – Trump, Berlusconi, the Shinawatras in Thailand. The trick is often to project an image of economic success coupled with distaste for the character, the lures and the corruption of establishment politicians. Mudde and Rovira Kaltwasser quote Berlusconi again: 'I don't need to go into office for the power. I have houses all over the world, stupen-dous boats . . . beautiful airplanes, a beautiful wife, a beautiful family . . . I am making a sacrifice.'[22]

At the time of the Brexit referendum, a good friend of mine – a cheerful, generous, larger-than-life working-class woman – told me she was voting for Boris Johnson 'because he's like me'. Boris Johnson – privileged, wealthy, unashamedly dishonest and self-serving – somehow, 'just like me', and somehow, 'on her side'.

When it is a matter of us and them, the clever leader imitates the very features of 'us' that draw criticism from 'them'. In the culture wars of modern America, the woke ideas of the liberal left leave many people keeping quiet about thoughts and feelings that do not match this new cultural imperative. But nobody likes to be told what to feel and think, and perhaps Trump's appeal, at least in part, arises from the self-absorption, abuse of others and general small-mindedness put clearly on display. Selfish, abusive or small-minded thoughts may not be pretty, but they are also not unknown – I suspect, to any of us – and perhaps seeing these things in a leader makes our own vices more tolerable.

Pan writ large: Competing IRMs

In the Pan of political choices, there is more than 'us' and 'them', and more than the charismatic appeal of a strong leader. In Chapter 7, I discussed Jonathan Haidt's idea of 'moral foundations' such as care/harm and sanctity/degradation. In Haidt's thinking, each foundation is a universal in human moral values, built on its own set of IRMs filled in with the specifics of an individual culture. Each one, Haidt argues, can be used as the basis for political appeal.

In *The Righteous Mind*, Haidt presents the results from an online survey of more than 130,000 people.[23] Each person's answers allowed them to be placed along the political continuum, from very liberal to very conservative and, at the same time, indicated their strength of concern for each of Haidt's moral foundations. The results tell us much about the different values of the left and the right. Concern for the care/harm foundation decreases systematically from very liberal to very conservative, directly in line with liberals' belief that society should protect the needy. The same is true for fairness/cheating. For other values, however, it is the conservative who is more concerned. These include loyalty/betrayal – reflecting commitment to the 'us' – sanctity/degradation and authority/subversion. Again, we are faced with multiple IRMs, sometimes directly in conflict. Haidt argues that their different weights have much to do with our political preferences, and the leaders and policies we support.

In all of this there is very little Spock, in the form of rational policies designed to better the people's lot. In turn, as Jonathan Haidt argues extensively in *The Righteous Mind*, it may be hopeless to fight the IRMs of leadership with appeals to reason. This is simply not the battle that a leader like Trump is fighting and winning. Obvious lies, disorganisation and inability to form effective policy and plans simply do not matter, because the leader has managed to make the battle all about Pan. Perhaps, as Haidt argues, it is more effective to fight Pan with Pan. Discussing the aggressive power of laughter, Lorenz notes its use in debunking the false enthusiasm of the manipulating leader, in words that could be a general-purpose scripting rule for political satire:

> Humour is the best of lie-detectors and it discovers, with an uncanny flair, the speciousness of contrived ideals and the insincerity of simulated enthusiasm. There are few things in the world so irresistibly comic as the sudden unmasking of this sort of pretence. When pompousness is abruptly debunked, when the balloon of puffed-up arrogance is pricked by humour and bursts with a loud report, we can indulge in uninhibited refreshing laughter which is liberated by this special kind of sudden relief of tension.[24]

In *The Righteous Mind*, Haidt describes how, during the George W. Bush–John Kerry presidential election, he spent long evenings walking his dog and mentally rewriting the ineffectual Democratic campaign messages to better match the releasing conditions of his moral foundations. Instead of 'America can do better', Haidt in his mind picked on each empty Bush promise and asked, 'You gonna pay for that, George?' – using the fairness/

cheating foundation to unmask the impractical combination of expensive programmes and radical tax cuts.[25]

The echo chamber

Any political commentator will nowadays speak of the 'echo chamber', and the increasing power of the internet to serve up only material that reinforces a pre-existing view. Modern algorithms deliver material that not only matches a political position but strengthens it, often out of all proportion. These algorithms want the reader's attention and, very soon, they discover that the more extreme the material, the greater the attention it receives.

In *The Chaos Machine* (2022), the journalist Max Fisher links this newly created power to deadly world events – the Rohingya genocide in Myanmar and a Zika epidemic in Brazil fuelled by anti-vax parents. Sometimes the echo chamber is created deliberately, in news stations and information outlets that promote an explicit agenda. Sometimes it just happens, as algorithms designed to attract advertising revenue pull the user further and further into some rabbit hole of one-sided belief.

In my book *How Intelligence Happens*, I described the experience of an old friend of mine, a social studies teacher in a Colorado high school. She told me that:

> in social studies class, children really love the political debates . . . for one of the few times in modern American life, they get to discuss real issues, not with pundits, pastors, or parents, but with an actual cross-section of normal, real people with their own, individual minds. With a newspaper or a television program, we perhaps have too much control . . .

if the arguments do not appeal, we can press a button and change the channel. Perhaps this makes it all too easy to stay trapped in our own, one-sided mental creations. Especially when they do not agree with us, perhaps one great advantage of real, other people is that they are not so easy to turn off.[26]

Hearing this, it is impossible not to imagine the merits of democracy in the Golden Age of Athens, with at least 6,000 people crammed into the debate, and any citizen free to speak up.

The laws of the human mind exaggerate the echo chamber effect, with our well-known preference for information that matches our preconceptions. Here is a nice example. In these experiments, the psychologists Vincent Frigo, Lang Chen and Timothy Rogers chose a task as emotionally neutral as possible, as far as they could get from preconceptions or social pressures.[27] Participants imagined that they were shipwrecked on a desert island, and had to learn whether fruits they saw were better eaten cooked or raw. On each trial of the experiment, the participant saw the picture of a fruit, along with advice from one or two other crew members giving their own opinions. At the start, the participant was trained in an initial belief, which they knew to be imperfect but a reasonable starting point – they also knew that the opinions they were given by their companions could be useful but were also imperfect. By measuring how the participant's responses changed through the experiment, the research team could assess whose opinions were listened to. If there was just a single companion, his or her opinion was always used – over trials, the participant's own belief shifted in the direction of the advice they were given. The key results came when there were two companions, sometimes giving rival advice. Both sets of advice had some effect, but they were differentially weighted. The companion whose beliefs were closer to the participant's was believed more. The companion whose beliefs were far away tended to be disregarded. Most important of all, if the beliefs of one companion were very close to the beliefs that the participant held already, then the other, dissenting voice was disregarded altogether.

As the authors of this study argue, just one voice strongly supporting a preconception may be all it takes to keep this preconception in place, entirely discrediting any rivals.

Ideal policy

The ideal political leader is acting for the best of the community. Formulating policy requires complex systems for gathering and integrating information, passed on to the government for final decision. Obviously, any political system has a great deal of this but, equally obviously, it is not the only force at play in the leader's mind. Along with the Spock of predicting and controlling world events, there is the much simpler Pan of any human being – the urge to succeed, to rise in the pecking order, to be loved and admired, and often, to gain privilege and luxury.

News images of political leaders only rarely show real joy when a new policy is unveiled, or even when later evidence shows it is a success. Joy is what we see when the candidate wins the election, with fists pumping the air and triumphant hugs. The politician is like anybody else, with Spock and Pan each contributing their prompts and needs. It is not too cynical to imagine that, for many politicians, the desire to be elected and to win often outstrips the desire to benefit the people's lot. Indeed, the whole point of democracy is its intention to rein in the urge of individual leaders to pursue personal gain. In ancient Athens, an excess of luxury or ostentation was considered undemocratic, a suggestion of 'Persian leanings'. In the temple to Apollo at Delphi, along with the inscription 'Know thyself' was 'Nothing in excess'.

We should like leaders who weight the good of the people over their own need for success, but as any glance at the news will show, elections are very far from a guarantee that we will get them. Of course, we are right to think that the leaders and the political elite cannot be trusted to have our own interests closest to their hearts. When a leader is chosen, it should be the leader whose policies bring most benefit to the people. For democracy to work best, we need voters informed by the best possible knowledge of what leaders and their policies will achieve.

Voters do not have access to the infinite knowledge of a flawless Spock, but it seems highly likely that, as the power of artificial intelligence leaps forward, there will be rapid improvements in the knowledge that they do have. In August 2023, the BBC published a story headed, 'Netflix: How did it know I was bi before I did?'[28] Written by BBC reporter Ellie House, it was the now-familiar story:

> I'd had one long-term boyfriend before then, and always considered myself straight. To be honest, dating wasn't at the top of my agenda. However, at that time I was watching a lot of Netflix and I was getting more and more recommendations for series with lesbian storylines, or bi characters. These were TV series that my friends – people of a similar age, with a similar background, and similar streaming histories – were not being recommended, and had never heard of.

Ellie House had not come out to herself – but by her choices of what to watch, she had come out to Netflix. We like to think we are complicated and sometimes private, but by everything that we do, all day every day, we leave a visible fingerprint of who we are and it is no longer science fiction for artificial intelligence to read these fingerprints.

Politicians, like anybody else, are going to vary in the weighting they give to different priorities and goals. It is easy enough to deduce that Johnson and Trump care mainly about winning, and much less about doing good for the people. For most politicians, perhaps, it is harder to say. The fingerprint is there, however, in everything they have ever written or said, and every policy they have formulated or supported. It would not surprise me if, already, systems are being built to read this fingerprint and tell us what sort of person we are really dealing with. For the voter, such a system would be like getting a synthesis of advice from everybody the candidate has ever known or influenced. Pan makes it all too easy to ignore even the best advice, but still – informed advice is informed advice, and usually we are better with it than without it.

More speculatively, the future of artificial intelligence could offer a good bit more than this. By the time anybody reads my words, artificial intelligence will have leaped yet further forward in its ability to take in and synthesise knowledge, understanding and predicting events in the world. It may be impossible to know the future of this even just a few years in advance, but whatever that future is, it will be a matter of ever-increasing power in acquiring and synthesising complex data. I find it very easy to imagine that, some time fairly soon, the best artificial systems will outperform the best-informed economists and planners in predicting the consequences of a policy. As these systems become increasingly available, which surely they will, it could become increasingly difficult for a politician to propose the same sort of dangerous policies that they are free to propose today – designed not because they are a good idea for the people's well-being, but because they appeal to Pan.

In this respect, we could be approaching an era of more informed democracy, with high-quality expertise available even to the average citizen, and a reducing need to trust elites and leaders. In our complex political decisions, perhaps this can help shift the balance of power from the simple urges of Pan to the informed deliberation of Spock.

Spock and Pan: Formulating policy and winning support

The best intentioned government is limited by what the people will agree to, and in all people, there are the twin voices of Spock and Pan. We do want the government to make our lives better, but with all the impossibility of really thinking it through, we fall heavily back on Pan – the real or fictitious enemy, the enormous poster of a severe leader, the appeal to simple values and traditions. Even a well-informed and well-intentioned government cannot ignore this. Mudde and Rovira Kaltwasser put it like this: 'In many western European countries the established parties have prioritized responsibility over representation [with a] consequent loss of public support.'[29] In democracy, even when the government sincerely does its best, there is no guarantee that the people will like it.

There is no guarantee – but here, surely, is the real answer to why democracy is the best form of government. With party manifestos, election promises and attractive-sounding policies, democracy focuses on its positive side. We see it as a process allowing the people to choose the leaders and the future they want. But here is its obvious weakness – to formulate and choose policy, the will of the people is simply too unstable, too uninformed and, above all, too Pan. Instead, perhaps, the real importance of democracy lies in its negative side – in its ability to restrict the power of leaders who fail to deliver. They may fail because the task is too difficult, or because they are incompetent, or because they are corrupt and self-serving. Without democracy, their incompetence or greed can run unchecked, but with an electorate to judge them, they can be replaced. Even in this negative sense, it is obvious that elections do not work perfectly – but surely we are better with them than without them.

Epilogue

Balancing

A good life and the rules of the mind

How should we live our lives? We want a life that is productive, decent and civilised, and at the same time rewarding, important, fulfilling. We want to live our lives well – but what is a good life? When we die, will we know whether we lived it?

I have suggested that the answers of great thinkers, from the ancient Greeks to the present, are not right. One thought is that human life should be guided by reason, but reason alone, I have argued, can never be enough. Another is that the answers come from outside ourselves, perhaps from God, but I think the answers are our own, supplied by the very way our own minds are built. If we understand ourselves, and the very general principles of our minds, then we understand the conflicts and paradox of human life and our human needs. To know a good life, we must know ourselves.

There are two sides to ourselves, and their principles are very different. One is the animal side, personified by Pan, and its principles are the principles of animal behaviour and the innate releasing mechanism. The IRM is enclosed and often short term, dictating the particular inner or outer response to particular environmental events. IRMs can be simple, such as the response of delight to our child's smile, or elaborate, such as the need to respond to a gift with one equally matched, or the need of a man facing court martial to talk it over with his wife. IRMs are in conflict, like the IRMs of a fish wishing simultaneously to approach and to flee, or a girl photographed by Eibl-Eibesfeldt, looking up towards the photographer with a shy smile, then away, then back again. IRMs are inconsistent, their power waxing and waning with time, context or changes in the season. Above all, IRMs need to be discharged, and when we discharge them, we have a sense that life is meaningful and right, a sense that we are fulfilled. It can be care for a neighbour, or defeat of an enemy, or intense sexual gratification, or struggle to the hilltop. It is these things that feel as though they deeply matter – not the spiritual calls of the gods, but the animal calls of ourselves.

The idea side is something quite different. Now, any mental contents, anything we can conceive or imagine, can be sewn together into any new idea. It can be an idea about geometry, or particle physics, or the law, or a religious edict. It can be a great idea, or a bad one. When good ideas are strung carefully together, we have 'reasoning', taking us from things we already knew to new, often amazing conclusions. When ideas are simply generated, often developing something prompted by Pan, they can lead us forward or astray. Civilisation would be impossible without Spock, just as medicine or space travel would be impossible, but Spock can also trap us, with an over-simplified, unreal and joyless take on life. Spock must tell us 'Thou shalt not kill' and demand that, in our civilisation, men and women have equal rights and equal opportunity to realise their dreams. But Spock may also specify the restrictive rules of a harsh religious observance; or move from equity to gender feminism, and the belief that women not only can but must struggle with the men in the fields of the kibbutz; or tell us that riches bring contentment, when the deeper contentment comes from the struggle and success that earned riches as a reward.

Often we look down on or fear our animal side, but the animal side is not all the savage needs of Mr Hyde. It gives us love as well as hate, the joys of the family, the support of friends and the thrill of effort and achievement. As Viktor Frankl thought, friends, struggles, achievements bring meaning to life. Just as David Hume argued, furthermore, there can be no reason without a starting point, and Pan is essential in the starting points he provides. Spock's ideas build religious, legal and moral rules, but the starting point is provided by Pan's social instincts – the instincts for mutual care, fairness, exchange, loyalty. Just as our individual culture fills the instinct to express our thoughts with the words and structure of our own language, Spock fills our instinctive moral world with the particular rules of our own, individual culture.

This image of ourselves shows that we are not single, coherent wholes. Instead, there is the patchwork – many promptings from Pan, many ideas generated by Spock, all calling to control our thoughts and our lives. Sometimes, despite their very different origins, Pan and Spock may agree; very often there is conflict. In his parliament of the instincts, Lorenz envisioned many IRMs, varying up and down in their weight, competing to control the animal's behaviour. In our own world, the complexity is increased many times over, with our own complex IRMs, varying up and down in weight, varying from one person to another, and then a potentially infinite set of ideas, some congruent, some incongruent, with salience varying over time and circumstance, and with each culture and each person having ideas of their own. The principles may be general, but filled in with the details of an individual person living an individual life, the result is as unpredictable as the falling of an individual leaf, governed by Newton's laws but played upon by an unmeasurable constellation of forces.

These rules of the human mind are everywhere. They shape our beliefs, our choices, our laws, our culture, every corner of our lives. Psychology is sometimes thought of as 'soft science', but these principles are not soft. With them, we understand ourselves.

And – it matters. Both Pan and Spock are essential to our lives. Both Pan and Spock can lead us hopelessly astray. We need to be open to their very different strengths and needs. We need to avoid their very different mistakes. With the strengths and weaknesses of our two different selves, we are always balancing.

Meaning in today's world

If Pan brings our sense of meaning and fulfilment, it is easy to imagine that, in the modern, anonymous world, Pan all too often is lost. As I have argued earlier, we do not understand ourselves well and, without realising the cost, we end up with countless hours spent in activities that our modern world demands, but that leave Pan unmoved. The commuting, the tax returns, Rutger Bregman's bullshit jobs, the boxes ticked in endless paperwork. It expands, it cannot easily be avoided, but Pan honestly, really doesn't care.

At the same time, what Pan does care about is challenged by a world whose rules are not his rules. The children grow up . . . and are gone. Lifetime friends are not lifetime friends at all . . . they now work far away and, over the years, friendship has degenerated to Christmas cards. Surely the sense of achievement in the great majority of modern jobs is nothing like the sense of achievement for a !Kung hunter, distributing meat to the hungry community. When I come home from the supermarket, I like to enter the house shouting, 'I've been hunting!' It's something, but it's really not like those summer dusks from my childhood, out with my father checking the rabbit traps.

While Pan stays still, the world changes around him at quite a bewildering speed. In just 10 to 15 years, the smartphone has changed everything. It is now quite hard to remember what crowded public spaces were like at the turn of the century – perhaps an airport waiting area, filled with people chatting, hugging, checking out the crowd for an attractive or beguiling stranger. A person growing up today will perhaps never know a situation like this because, for better or worse, the phone is simply more interesting than the people around us. On the one hand, the phone offers Pan something new. The children may live far away, but now all day there is the option of a quick message and a feeling of contact. On the other hand, Pan was designed for real, physical contact, and the phone will not offer a touch of hands or a hug. When I see a modern airport, I simply marvel at the flexibility of the human mind. In a few years, everything it was designed for has changed again – all those social IRMs, all those fixed action patterns are now discharged at a small screen. Yet it does work, at least reasonably well and often extremely well. We can push Pan to a quite staggering degree, but he is still there, demanding to be satisfied.

Every argument has two sides

In the jungle of competing human values and needs, every choice of how we should act has costs as well as benefits. We have seen many examples – the competitive striving in the workplace, that both drives activity forward and inhibits the less competitive colleague; the gaze at the opera, that can be received either as harassment or as opportunity; the aggression of 'laughing at', often hilarious for those who laugh but diminishing for the victim. With his tendency to focus, Spock struggles to grasp just how partial his current thought is. We believe quite passionately in our own culture's ideals – in our views about decency, human rights, the way to raise our children. The passion is important to us, but perhaps we can remember that, in the complex world of human society, our deeply held beliefs are never the whole truth. As our culture changes, with the speed of Spock's new ideas and their rapid spread through the population, of course there will be culture wars, with some seeing the costs and some the benefits. As I have said, these are wars with one another, but also wars between different sides of ourselves.

Faced with these wars, 'Know thyself' again may not be a bad maxim. As Spock turns reality into a set of ideas, reality is the same for people on opposite sides of a cultural divide, but the ideas can be quite different. Christian and Muslim can passionately believe that their own version of reality is right, but surely it must help to reflect that the particular ideas are an accident of birthplace, and whatever reality there is, it must be reality common to both these two religions and any others. With his penchant for over-simplification, Spock turns neglect of women's rights into a vigorous gender feminism, seeing everything in our culture through a lens of oppressive patriarchy. But a sensible and fulfilled life cannot be built on a war of 4 billion women against 4 billion men, each one a complex, multi-dimensional self. As one idea more than another dominates our thinking, it is inevitable that we see only a part of the picture, and realising this, we can perhaps be more open to the other parts seen by other people.

Our ideas are not the whole truth, and if they are implemented, the results can be not at all what we imagined. There is the kibbutz movement, which did not at all achieve the abolition of motherhood – but did, perhaps, achieve a sense of equal value between the sexes. From Leda Cosmides and John Tooby, there are the examples of rent control and Marxist agriculture, and the negative effects of their implementation. In the workplace, to promote the equality of men and women, we may suppress 'toxic' masculine competition – but, as I have argued, the joy of battle also has its upside in driving forward human effort, and it is hard to know what we may lose when we suppress it and the work itself, quite possibly, loses one aspect of its glamour. In all these cases, we may passionately believe in our cultural choices, but as we put them into practice, very likely we are in for surprises.

A life plan – with the right building blocks

My friends and family would find it rather amusing to think of me, of all people, writing a book about the need to set Pan free. Pan is short term, directing the particular response to a particular situation. A nesting cormorant returning with food for the chicks is not thinking about the life that these chicks will lead as adults – it just brings the food – and our Pan doubtless is just the same, responding to the sight of our own crying child, or to a gift from a neighbour. A human life, however, needs a long-term plan, with relationships, career plans, achievements built over many years. Few people are more about the plan than I am and, surely, the plan must be Spock's creation, with his thoughts not just about today, but about the whole lifetime ahead.

Our plans cannot constantly be derailed by the IRM of the moment – the inviting look of a potential new sexual partner, or the lure of a day abandoning work to win at the races. But as we construct the plan, we need to use the building blocks we have inside us – the needs, the behaviour patterns and the fulfilment of Pan. As we delight in playing with our child, we play, as Buddhists say, 'in the moment', for now knowing nothing but the child's thrilled laughter. Later on, we can worry about gym kit for the next day, or the impending school exams. The same as our partner lights a candle by the bedside – right now we let Pan free, lost in one another's bodies, and for now this is not a promise to be there over the decades of building a family and a life. Pan on his own cannot possibly be enough, but without Pan's blocks to build with, Spock's plan is empty. As I have said throughout this book, it is a matter of balance, of weaving the two threads of ourselves together.

Again, it is 'Know thyself'. Though Pan is no Mr Hyde, his strength and vitality obviously can be dangerous. We do in the moment long to pursue that attractive young woman or man, or to show off to our colleagues when it would be much gentler to keep quiet, or to scream at the motorist who cuts us off. But Pan's fixed behaviour patterns, and Pan's urge to be discharged, are relatively easy to understand and control. As Lorenz says, informed by Spock's understanding of himself, the man whose aggressions have built to fever pitch goes outside and beats up some hapless inanimate object. The couple building a life together can choose simple things – hidden favours, shared achievements, shared music – which just like the greylag's triumph ceremony, really do cement the bond. Pan provides vitality; Spock provides understanding and stability to keep life on the rails.

Does culture advance?

Thinking of Roman slaves and crucifixions, or reading of beaten schoolchildren in D.H. Lawrence's *The Rainbow* (1915), it is impossible not to feel that we have, in some real sense, moved forward. We feel irresistibly that culture 'advances'. As I said earlier, though, I do not think I could prove to a slave captain that he was mistaken – I could just tell him that I will not tolerate slavery. To say that things have got better, we need a scale determining what is better or worse, and in our jumble of Pan and Spock, there is no single master scale, only multiple, sometimes competing values. The value of communal sharing may say that our possessions should be given away; the value of ownership and the right to keep what we have earned may say they should not. The slaver and I weight human values differently, and I may strongly prefer my weights – but to decide how weights *ought* to be set, we would first need to agree on some scale for comparing the merits of different weighting schemes, and no objective basis for such agreement seems possible.

A similar indeterminacy arises with every new cultural thought – perhaps that we should not shame people for their body shape, or not tell jokes about nagging wives or their blinkered husbands. We can see the reason for such thoughts, and it is tempting to say that our revised culture just got better. But it also got worse – a few million people thoroughly enjoyed celebrating how their wives badger them or their husbands are idiots, and now they can't. Of course, they resent the loss and, if they wish, they can see the loss as more important than the gain.

Perhaps, at least, we can agree that the abolition of slavery and school beatings makes culture more civilised. Modern, large-scale societies demand respect for the rights of others, even an immense body of anonymous strangers, and when we decide not to take slaves or to allow schoolteachers to pick up a cane, we are making another move towards equality for all. This may not be the only possible human value, but it is one that our large-scale societies depend on. In this sense, there is 'progress' – but it should never surprise us when, with different values in mind, a different person sees things differently.

As regards political correctness and our modern culture wars,

I have argued that, great though it is, our determination to treat all people as equivalent just cannot work. We deeply need the sense of ourselves, of belonging within our own personal history and our own individual community and culture. As we struggle to live well within our own personal values, we need the esteem of ourselves and of the community we value. People are not faceless equivalents, like a horde of ants or bees, or the marching human hordes in Yevgeny Zamyatin's 1920s novel *We*, and when Spock denies this, human life is diminished. Civilised culture needs a way to combine the equivalent rights of all with our irresistible need to be ourselves.

The mystery of meaning

In everything I have said there is a deep question. Following Lorenz and Frankl, I have argued that it is our animal side – the discharge of our human IRMs – that gives life its sense of right-ness and meaning. But how can this be? IRMs are small bits of computer code, often wired into the most ancient structures of the brain, and sometimes dependent on just a small group of neu-rons, perhaps in some small nucleus of the hypothalamus. Why would it be – indeed, how on earth *can* it be – that when these neurons fire, we feel that our life is right, but when millions or billions of other neurons fire, the result is just another idea?

The feeling of meaning is very special to us – it is perhaps something that we think could never happen in an artifi-cial agent, no matter how complex and intelligent that agent becomes. But obviously, the feeling has nothing to do with com-plexity. It would actually be very simple to program an artificial agent to peck at the red dot on a parent gull's beak, or to pick up a crying baby and jiggle it around until it smiles. By the standards of mental life, these are not complex things – they are very simple things, and just a small bit of code would be enough to make them happen. But when that code runs in our own brains, something very special happens. Like other conscious experiences, this one is somehow a property of a particular electrical event in our brain – and one that in this case means the world to us.

If our own sense of meaning is built so strongly into our animal side, generated by neural activity in ancient brain systems that we share with many other animals – what then of those other animals? *Drosophila* has IRMs. A mouse has IRMs which very often resemble ours and, I strongly suspect, are controlled by very similar neural events. Like ours, the IRMs of *Drosophila* and the mouse have their weights that vary over time and circumstance, and their own spontaneous need to be discharged. As the season to migrate approaches, a caged bird turns with increasing restlessness in the direction it should follow, the urge to go manifest in its straining against the walls that keep it captive. It is hard to imagine that, when the bird is released and it can at last take flight, it does not 'feel right'. But what on earth does that mean – in the bird, or in us?

Acknowledgements

I am more than grateful to my agent, Bill Hamilton, for believing in this project and guiding it to completion; and to Jamie Joseph and the expert staff at Ebury, for their efficiency and generosity in making the book better. I am also grateful to the friends and relations who have read parts and helped me to straighten out the ideas – sometimes by agreeing, often the opposite. They include Peter Cooper, Kyle and Pete Duncan, Marge Eldridge, Lynne Murray, Jane Raymond and Alex Woolgar.

Notes

INTRODUCTION

1. https://www.bbc.co.uk/news/resources/idt-e885e189-aee4-4250-bbfc-c7f3c52f4d48 (published 11 August 2024); https://www.bbc.co.uk/news/world-us-canada-65071989 (published 25 March 2023); https://www.bbc.co.uk/news/uk-65015479 (published 21 March 2023); https://www.bbc.co.uk/sport/football/61099003 (published 15 April 2022); https://www.bbc.co.uk/news/world-europe-60600487 (published 4 March 2022); https://www.bbc.co.uk/news/uk-england-derbyshire-5986904 (published 5 January 2022). The headlines are from the *BBC News* site; for each one I have added a summary of the accompanying story.

2. https://www.youtube.com/watch?v=Qv4-m-clZf4.

3. K. Duncker (1945), 'On problem solving', *Psychological Monographs* 58/5.

4. A. Newell, J.C. Shaw and H.A. Simon (1958), 'Elements of a theory of human problem solving', *Psychological Review* 65: pp. 151–66.

5. C. Darwin (1871/2013), *The Descent of Man* (Wordsworth Editions), p. 78.

6. K. Lorenz (1966/2002), *On Aggression* (Routledge), p. 46.

7. L. Cosmides and J. Tooby (1992), 'Cognitive adaptations for social exchange', in J.H. Barkow, L. Cosmides and J. Tooby (eds), *The Adapted Mind: Evolutionary Psychology and the Generation of Culture* (Oxford University Press).

8. D. Hume (1739), *A Treatise of Human Nature*, 2.3.3. See https://davidhume.org/texts/t/2/3/

9. I. Eibl-Eibesfeldt (1989), *Human Ethology* (Aldine de Gruyter), p. 608.

10. Saki, *Selected Stories by Saki* (Rupa Publications), p. 46.
11. Saki, *Selected Stories by Saki* (Rupa Publications), p. 91.
12. https://www.youtube.com/watch?v=D9Ihs24Izeg.

CHAPTER 1

1. N. Tinbergen (1972), *The Animal in Its World: Explorations of an Ethologist* (Harvard University Press).
2. K. von Frisch, *The Dancing Bees: An Account of the Life and Senses of the Honey Bee* (Harvest Books).
3. K. Lorenz (1970), *Studies in Animal and Human Behaviour* (Methuen), vol. 1, p. xvi.
4. D.R. Rubenstein and J. Alcock (2019), *Animal Behavior: International Eleventh Edition* (Oxford University Press), p. 104.
5. Rubenstein and Alcock (2019), p. 104.
6. Lorenz (1966/2002), p. 8.
7. K. Lorenz (1935), 'Der Kumpan in der Umwelt des Vogels. Der artgenosse als auslösendes moment sozialer verhaltungsweisen', *Journal für Ornithologie* 80: pp. 137–213.
8. Rubenstein and Alcock (2019), pp. 105–11.
9. A. Heyde, L. Guo, C. Jost et al. (2021), 'Self-organized biotectonics of termite nests', *Proceedings of the National Academy of Sciences USA* 118: e2006985118.
10. I. Eibl-Eibesfeldt (1971), *Love and Hate* (Methuen), pp. 104–6.
11. Eibl-Eibesfeldt (1971), pp. 10–11.
12. N. Tinbergen (1932) 'Über die Orientierung des Bienenwolfes (*Philanthus triangulum* Fabr.)', *Zeitschrift für Vergleichende Physiologie* 16: pp. 305–34.
13. Rubenstein and Alcock (2019), pp. 25–9.
14. J.S. Huxley (1923), 'Courtship activities in the red-throated diver (*Colymbus stellatus* Pontopp.): Together with a discussion of the

evolution of courtship in birds, *Zoological Journal of the Linnean Society* 35: pp. 253–92.

15. Lorenz (1966/2002), pp. 55–62.
16. Eibl-Eibesfeldt (1971), p. 44.
17. Eibl-Eibesfeldt (1971), p. 44.
18. Rubenstein and Alcock (2019), p. 287.
19. Examples taken from: Rubenstein and Alcock (2019), pp. 266–9; Lorenz (1966/2002), pp. 106–11; Eibl-Eibesfeldt (1971), pp. 64–6.
20. Lorenz (1935).
21. Lorenz (1970), p. 374.
22. Lorenz (1966/2002), p. 113.
23. N. Tinbergen and A.C. Perdeck (1950), 'On the stimulus situations releasing the begging response in the newly hatched herring gull (*Larus argentatus* Pont.)', *Behaviour* 3: pp. 1–39. See also Rubenstein and Alcock (2019), pp. 101–3.
24. Lorenz (1966/2002), chapter 6.
25. Quoted in Lorenz (1966/2002), p. 85.
26. Eibl-Eibesfeldt (1971), p. 107.
27. Rubenstein and Alcock (2019), p. 315.
28. Lorenz (1966/2002), pp. 93–9.
29. J.R.R. Stroop (1935), 'Studies of interference in serial verbal reactions', *Journal of Experimental Psychology* 18: pp. 643–62.
30. Lorenz (1966/2002), p. 49.
31. Eibl-Eibesfeldt (1971), p. 67.
32. Lorenz (1966/2002), pp. 49–50.
33. Rubenstein and Alcock (2019), p. 64.
34. Lorenz (1966/2002), p. 53.

CHAPTER 2

1. For discussion and examples, see: Eibl-Eibesfeldt (1989), pp. 60–1.

2. N. Tinbergen (1951), *The Study of Instinct* (Oxford University Press).

3. Eibl-Eibesfeldt (1989), p. 61.

4. See: Lorenz (1966/2002), pp. 260–1; Eibl-Eibesfeldt (1971), p. 16.

5. Eibl-Eibesfeldt (1989), p. 138.

6. Eibl-Eibesfeldt (1971), p. 110.

7. Eibl-Eibesfeldt (1989), p. 138.

8. P. Ekman (1992), 'An argument for basic emotions', *Cognition and Emotion* 6: pp. 169–200.

9. Eibl-Eibesfeldt (1971), p. 18.

10. Eibl-Eibesfeldt (1989), p. 474.

11. Eibl-Eibesfeldt (1989), pp. 137–8.

12. C. Darwin (1872/1999), *The Expression of the Emotions in Man and Animals* (Fontana), pp. 22–4.

13. Eibl-Eibesfeldt (1971), p. 12.

14. Eibl-Eibesfeldt (1989), p. 369.

15. Darwin (1872/1999), p. 348.

16. For this quote, experimental findings and a summary of Tomasello's thinking, see:
    M. Tomasello, M. Carpenter, J. Call et al. (2005), 'Understanding and sharing intentions: The origins of cultural cognition', *Behavioral and Brain Sciences* 28: pp. 675–91.

17. E. Herrmann, J. Call, M.V. Hernandez-Lloreda et al. (2007), 'Humans have evolved specialized skills of social cognition: The cultural intelligence hypothesis', *Science* 317: pp. 1360–6.

18. Eibl-Eibesfeldt (1989), p. 170.

19. For a discussion of human appeasement and the link to chimpanzees, see: Eibl-Eibesfeldt (1971), pp. 169–71.

20. Eibl-Eibesfeldt (1989), pp. 493–5.

21. Eibl-Eibesfeldt (1989), p. 498.

22. J. Panksepp and W.W. Beatty (1980), 'Social deprivation and play in rats', *Behavioral and Neural Biology* 30: pp. 197–206.

23. W.D. Hamilton (1996), 'Foreword', in S. Turillazzi and M.J.

West-Eberhard (eds), *Natural History and the Evolution of Paper Wasps* (Oxford University Press).

CHAPTER 3

1. E.C. Tolman (1948), 'Cognitive maps in rats and men', *Psychological Review* 55: pp. 189–208.
2. Tinbergen (1932).
3. E. Tulving (1972), 'Episodic and semantic memory', in E. Tulving and W. Donaldson (eds), *Organization of Memory* (Academic Press).
4. Newell, Shaw and Simon (1958), pp. 151–66.
5. J. Duncan (2010), *How Intelligence Happens* (Yale University Press).
6. J. Duncan, D. Chylinski, D.J. Mitchell et al. (2017), 'Complexity and compositionality in fluid intelligence', *Proceedings of the National Academy of Sciences USA* 114: pp. 5295–9.
7. Duncker (1945).
8. A. Luria (1966), *Higher Cortical Functions in Man* (Tavistock).
9. J. Duncan, H. Emslie, P. Williams et al. (1996), 'Intelligence and the frontal lobe: The organization of goal-directed behavior', *Cognitive Psychology* 30: pp. 257–303.
10. E. Shafir, I. Simonson and A. Tversky (1993), 'Reason-based choice', *Cognition* 49: pp. 11–36.
11. F. Strack, L. Martin and N. Schwarz (1988), 'Priming and communication: Social determinants of information use in judgments of life satisfaction', *European Journal of Social Psychology* 18: pp. 429–42.
12. D. Kahneman, A.B. Krueger, D. Schkade et al. (2006), 'Would you be happier if you were richer? A focusing illusion', *Science* 312: pp. 1908–10.
13. D. Schkade and D. Kahneman (1998), 'Does living in California make people happy? A focusing illusion in judgments of life satisfaction', *Psychological Science* 9: pp. 340–6.
14. Kahneman et al. (2006).

15. L. Festinger, H.W. Riecken and S. Schacter (1956), *When Prophecy Fails* (University of Minnesota Press).

16. J. Haidt (2012/2013), *The Righteous Mind* (Penguin).

17. https://www.bbc.co.uk/news/world-europe-60600487 (published 4 March 2022).

CHAPTER 4

1. D.H. Hubel and T.N. Wiesel (1979), 'Brain mechanisms of vision', *Scientific American* 241: pp. 15–63.

2. S. Dorkenwald, A. Matsliah, A.R. Sterling et al. (2024), 'Neuronal wiring diagram of an adult brain', *Nature* 634: pp. 124–38.

3. D.J. Anderson (2016), 'Circuit modules linking internal states and social behaviour in flies and mice', *Nature Reviews Neuroscience* 17: pp. 692–704.

4. E.D. Hoopfer (2016), 'Neural control of aggression in *Drosophila*', *Current Opinion in Neurobiology* 38: pp. 109–18.

5. C. Lenschow and S.Q. Lima (2020), 'In the mood for sex: Neural circuits for reproduction', *Current Opinion in Neurobiology* 60: pp. 155–68.

6. Anderson (2016); J.E. Lischinsky and D. Lin (2020), 'Neural mechanisms of aggression across species', *Nature Neuroscience* 23: pp. 1317–28.

7. J. Kohl, B.M. Babayan, N.D. Rubinstein et al. (2018), 'Functional circuit architecture underlying parental behaviour', *Nature* 556: pp. 326–31.

8. Lischinsky and Lin (2020).

9. Kohl et al. (2018); D. Wei, V. Talwar and D. Lin (2021), 'Neural circuits for social behaviors: Innate yet flexible', *Neuron* 109: pp. 1600–20.

10. Lischinsky and Lin (2020).

11. For reviews of the material in this section, see: Lischinsky and Lin (2020); Wei, Talwar and Lin (2021).

12. D. Mobbs, P. Petrovic, L.L. Marchant et al. (2007), 'When fear is near: Threat imminence elicits prefrontal–periaqueductal gray shifts in humans', *Science* 317: pp. 1079–83.

13. M. Saxe (2006), 'Uniquely human social cognition', *Current Opinion in Neurobiology* 16: pp. 235–9.

14. R. Rajimehr, H. Xu, A. Farahani et al. (in press), 'Functional architecture of cerebral cortex during naturalistic movie-watching', *Neuron*.

15. D.J. Mitchell, A.L.S. Mousley, M.A. Shafto et al. (2023), 'Neural contributions to reduced fluid intelligence across the adult lifespan', *Journal of Neuroscience* 43: pp. 293–307.

16. M. Assem, M.F. Glasser, D.C. Van Essen et al. (2020), 'A domain-general cognitive core defined in multimodally parcellated human cortex', *Cerebral Cortex* 30: pp. 4361–80.

17. Assem et al. (2020).

18. Luria (1966).

19. A. Woolgar, A. Parr, R. Cusack et al. (2010), 'Fluid intelligence loss linked to restricted regions of damage within frontal and parietal cortex', *Proceedings of the National Academy of Sciences USA* 107: pp. 14899–902.

20. E. De Falco, M. Ison, I. Fried et al. (2016), 'Long-term coding of personal and universal associations underlying the memory web in the human brain', *Nature Communications* 7: p. 13408.

21. J. Duncan (2001), 'An adaptive coding model of neural function in prefrontal cortex', *Nature Reviews Neuroscience* 2: pp. 820–9.

22. M. Rigotti, O. Barak, M.R. Warden et al. (2013), 'The importance of mixed selectivity in complex cognitive tasks', *Nature* 497: pp. 585–90.

23. A. Nieder (2017), 'Inside the corvid brain: Probing the physiology of cognition in crows', *Current Opinion in Behavioral Sciences* 16: pp. 8–14.

24. R. Desimone and J. Duncan (1995), 'Neural mechanisms of selective visual attention', *Annual Review of Neuroscience* 18: pp. 193–222.

CHAPTER 5

1. A. Norenzayan, A.F. Shariff, W.M. Gervais et al. (2016), 'The cultural evolution of prosocial religions', *Behavioral and Brain Sciences* 39: pp. 1–65.

2. H. Gintis (2011), 'Gene-culture coevolution and the nature of human sociality', *Philosophical Transactions of the Royal Society B* 366: pp. 878–88.

3. D. Sperber (1985), 'Anthropology and psychology: Towards an epidemiology of representations', *Man* 20: pp. 73–89.

CHAPTER 6

1. Eibl-Eibesfeldt (1971), p. 207.

2. L. Murray (2014), *The Psychology of Babies: How Relationships Support Development from Birth to Two* (Constable & Robinson).

3. M. Greenberg and N. Morris (2013), 'Engrossment: The newborn's influence upon the father', in S.H. Cath, A.R. Gurwitt and J.M. Ross et al., *Father and Child: Developmental and Clinical Perspectives* (Psychology Press).

4. Eibl-Eibesfeldt (1971), pp. 107–13.

5. Lorenz (1966/2002), p. 164.

6. Lorenz (1966/2002), pp. 169–70.

7. Eibl-Eibesfeldt (1971), p. 122.

8. Lorenz (1966/2002), pp. 177–208.

9. R.I.M. Dunbar (2018), 'The anatomy of friendship', *Trends in Cognitive Sciences* 22: pp. 32–51.

10. Dunbar (2018).

11. J. Lehmann, A.H. Korstjens and R.I. Dunbar (2007), 'Group size, grooming and social cohesion in primates', *Animal Behaviour* 74: pp. 1617–29.

12. https://www.youtube.com/watch?v=GJtq6OmD-_Y.

13. D.N. Stern, B. Beebe, J. Jaffe et al. (1977), 'The infant's stimulus world during social interaction: A study of caregiver behaviours with particular reference to repetition and timing', in H.R. Schaffer (ed.), *Studies on Interactions in Infancy* (Academic Press).

14. D.E. Brown (1991), *Human Universals* (McGraw-Hill).

15. V. Eleuteri, M. Henderson, A. Soldati et al. (2022), 'The form and

function of chimpanzee buttress drumming', *Animal Behaviour* 192: pp. 189–205.

16. H. Honing, F.L. Bouwer, L. Prado et al. (2018), 'Rhesus monkeys (Macaca mulatta) sense isochrony in rhythm, but not the beat: Additional support for the gradual audiomotor evolution hypothesis', *Frontiers in Neuroscience* 12: p. 475.

17. W.H. McNeill (1995), *Keeping Together in Time: Dance and Drill in Human History* (Harvard University Press).

18. R.I.M. Dunbar (2022), 'Virtual touch and the human social world', *Current Opinion in Behavioural Sciences* 43: pp. 14–9.

19. Eibl-Eibesfeldt (1989), p. 138.

20. R.R. Provine (2004), 'Laughing, tickling, and the evolution of speech and self', *Current Directions in Psychological Science* 13: pp. 215–8.

21. S.K. Scott, C.Q. Cai and A. Billing (2022), 'Robert Provine: the critical human importance of laughter, connections and contagion', *Philosophical Transactions of the Royal Society B* 377: p. 20210178.

22. https://www.ted.com/talks/sophie_scott_why_we_laugh?language=en.

23. S.K. Scott, N. Lavan, S. Chen et al. (2014), 'The social life of laughter', *Trends in Cognitive Sciences* 18: pp. 618–20.

24. Dunbar (2018).

25. https://en.wikipedia.org/wiki/Barn_raising.

26. B. Spock (1945/1979), *Baby and Child Care* (W H Allen), p. 361.

27. https://www.youtube.com/watch?v=2arlinLwqGQ.

28. J. van Lawick-Goodall (1968), 'The behaviour of free-living chimpanzees in the Gombe Stream Reserve', *Animal Behaviour Monographs* 1: pp. 161–311.

29. Eibl-Eibesfeldt (1989), p. 341.

30. Eibl-Eibesfeldt (1971), p. 183.

31. Eibl-Eibesfeldt (1971), p. 183.

32. B. Malinowski (1922), *Argonauts of the Western Pacific* (Dutton).

33. Eibl-Eibesfeldt (1989), p. 308.

34. Dunbar (2018).

35. Lorenz (1966/2002), pp. 249–59.

36. S. Freud (1921/1955), 'Group psychology and the analysis of the ego',
    in J. Strachey (ed.), *Standard Edition of the Complete Works of Sigmund Freud* (Hogarth Press), vol. 14.
37. Eibl-Eibesfeldt (1971), p. 196.

CHAPTER 7

1. A.P. Fiske (1992), 'The four elementary forms of sociality: Framework
   for a unified theory of social relations', *Psychological Review* 99:
   pp. 689–723; Cosmides and Tooby (1992).
2. Fiske (1992).
3. Fiske (1992), p. 697.
4. https://www.youtube.com/watch?v=QfgVhEiM6ns.
5. Fiske (1992), p. 693.
6. Fiske (1992), p. 697.
7. Fiske (1992), p. 697.
8. M.A. Hogg (2016), *Social Identity Theory* (Springer International).
9. Fiske (1992), pp. 697–8.
10. Fiske (1992), p. 700.
11. Eibl-Eibesfeldt (1989), p. 299.
12. See: Eibl-Eibesfeldt (1989), p. 443; G. Zivin (1977), 'Facial gestures
    predict preschoolers' encounter outcomes', *Social Science Information*
    16: pp. 715–30.
13. Lorenz (1966/2002), pp. 41–2.
14. Lorenz (1966/2002), p. 43.
15. H. Ford (1922), *My Life and Work* (Doubleday).
16. Fiske (1992), p. 701.
17. Eibl-Eibesfeldt (1971), p. 86.
18. Eibl-Eibesfeldt (1989), p. 306.
19. Fiske (1992), pp. 700–1.
20. Eibl-Eibesfeldt (1989), p. 305.

21. Fiske (1992), p. 704.
22. J. Henrich, R. Boyd, S. Bowles et al. (2005), '"Economic man" in cross-cultural perspective: Behavioral experiments in 15 small-scale societies', *Behavioral and Brain Sciences* 28: pp. 795–815.
23. S.F. Brosnan and F.B. de Waal (2014), 'Evolution of responses to (un) fairness', *Science* 346: p. 1251776.
24. Haidt (2012/2013).
25. Eibl-Eibesfeldt (1989), p. 344.
26. Darwin (1871/2013), p. 71.
27. R.I. Dunbar (2004), 'Gossip in evolutionary perspective', *Review of General Psychology* 8: pp. 100–10.

CHAPTER 8

1. Rubenstein and Alcock (2019), p. 213.
2. Eibl-Eibesfeldt (1989), p. 108.
3. Eibl-Eibesfeldt (1989), p. 81.
4. Darwin (1871/2013), pp. 71–2.
5. Lorenz (1966/2002), p. 243.
6. S. Pinker (2002/2019), *The Blank Slate* (Penguin), p. 57.
7. Eibl-Eibesfeldt (1989), p. 417.
8. Eibl-Eibesfeldt (1989), p. 417.
9. H. Tajfel (1970), 'Experiments in intergroup discrimination', *Scientific American* 223: pp. 96–103.
10. Herodotus (1972), *The Histories* (Penguin), p. 219.
11. https://www.youtube.com/watch?v=X1zFnyEe3nE.
12. E.H. Erikson (1985), 'Pseudospeciation in the nuclear age', *Political Psychology* 6: pp. 213–17.
13. Lorenz (1966/2002), p. 76.
14. Lorenz (1966/2002), pp. 77–8.
15. Eibl-Eibesfeldt (1989), p. 507.

16. Eibl-Eibesfeldt (1989), p. 403.

17. G. Hodson and K. Costello (2007), 'Interpersonal disgust, ideological orientations, and dehumanisation as predictors of intergroup attitudes', *Psychological Science* 18: pp. 691–8.

18. Eibl-Eibesfeldt (1971), pp. 94–5.

19. Eibl-Eibesfeldt (1971), p. 81.

20. T.F. Pettigrew and L.R. Tropp (2006), 'A meta-analytic test of intergroup contact theory', *Journal of Personality and Social Psychology* 90: pp. 751–83.

21. Pettigrew and Tropp (2006).

22. Lorenz (1966/2002), p. 274.

23. Lorenz (1966/2002), p. 283.

24. M. Sherif and C.W. Sherif (1969), *Social Psychology* (Harper and Row).

25. E. Viding and E. McCrory (2019), 'Towards understanding atypical social affiliation in psychopathy', *Lancet Psychiatry* 6: pp. 437–44.

26. E. O'Nions, C.F. Lima, S.K. Scott et al. (2017), 'Reduced laughter contagion in boys at risk for psychopathy', *Current Biology* 27: pp. 3049–55.

27. Viding and E. McCrory (2019).

28. T. Yang, C.F. Yang, M.D. Chizari et al. (2017), 'Social control of hypothalamus-mediated male aggression', *Neuron* 95: pp. 955–70.

29. Lorenz (1966/2002), p. 184.

30. Lorenz (1966/2002), p. 131.

31. S.J. Blakemore (2018), 'Avoiding social risk in adolescence', *Current Directions in Psychological Science* 27: pp. 116–22.

32. Eibl-Eibesfeldt (1989), p. 420.

33. Eibl-Eibesfeldt (1989), p. 604.

34. Haidt (2012/2013), p. 164.

35. Lorenz (1966/2002), p. 260.

36. De Bernières (2020/2021), *The Autumn of the Ace* (Vintage), p. 245.

37. Darwin (1871/2013), p. 73.

38. C.J. Vallgren (1996/2008), *Documents Concerning Rubashov the Gambler* (Vintage), p. 262.

CHAPTER 9

1. Lorenz (1966/2002), p. 234.
2. Eibl-Eibesfeldt (1971), p. 97.
3. Eibl-Eibesfeldt (1989), pp. 410–4.
4. Eibl-Eibesfeldt (1971), pp. 218–9.
5. Norenzayan et al. (2016), pp. 1–65.
6. Quoted in Haidt (2012/2013), p. 308.
7. Pinker (2002/2019), pp. 270, 274.
8. D. Hume (1748), *An Enquiry Concerning Human Understanding*, 1.14na. See https://davidhume.org/texts/e/iv.
9. Lorenz (1966/2002), p. 239.
10. Darwin (1871/2013), p. 71.
11. Darwin (1871/2013), pp. 66–7.
12. Pinker (2002/2019), p. 275.

CHAPTER 10

1. Quoted in R. Wacks (2006/2014), *Philosophy of Law* (Oxford University Press), p. 3.
2. D. Sznycer and C. Patrick (2020), 'The origins of criminal law', *Nature Human Behaviour* 4: pp. 506–16.
3. J. Rawls (1973), *A Theory of Justice* (Oxford University Press).
4. Haidt (2012/2013).
5. J. Griffin (2008), *On Human Rights* (Oxford University Press).
6. S.J. Gould (1998), 'The Diet of Worms and the defenestration of Prague', *Leonardo's Mountain of Clams and the Diet of Worms: Essays in Natural History* (Harmony Books).
7. Quoted in: Eibl-Eibesfeldt (1989), p. 710.
8. J.D. Greene, L.E. Nystrom, A.D. Engell et al. (2004), 'The neural bases of cognitive conflict and control in moral judgment', *Neuron* 44: pp. 389–400.

9. L. Cosmides and J. Tooby (2006), 'Evolutionary psychology, moral heuristics, and the law', in G. Gigerenzer and C. Engel (eds), *Heuristics and the Law* (MIT Press).

10. Cosmides and Tooby (2006), p. 201.

CHAPTER II

1. Darwin (1871/2013), p. 66.

2. Quoted in: Pinker (2002/2019), p. 297.

3. Eibl-Eibesfeldt (1989), p. 361.

4. C. Harmon-Jones, B.J. Schmeichel, E. Mennitt et al. (2011), 'The expression of determination: similarities between anger and approach-related positive affect', *Journal of Personality and Social Psychology* 100: pp. 172–81.

5. M. Lewis, M.W. Sullivan and H.M.S. Kim (2015), 'Infant approach and withdrawal in response to a goal blockage: Its antecedent causes and its effect on toddler persistence', *Developmental Psychology* 51: pp. 1553–63.

6. https://www.youtube.com/watch?v=HtTUsOKjWyQ.

7. https://www.youtube.com/watch?v=CSKiD3bZhRs.

8. Lorenz (1966/2002), p. 269.

9. V. Frankl (1959/2004), *Man's Search for Meaning* (Rider), pp. 82–3.

10. M. Rosenberg (1987), 'Rosenberg self-esteem scale', in K. Corcoran and J. Fischer (eds), *Measures for Clinical Practice* (Free Press).

11. R.F. Baumeister, J.D. Campbell, J. I. Krueger et al. (2003), 'Does high self-esteem cause better performance, interpersonal success, happiness, or healthier lifestyles?', *Psychological Science in the Public Interest* 4: pp. 1–44.

12. A. Maslow (1954), *Motivation and Personality* (Harper and Bros).

13. D.C. McClelland (1985), *Human Motivation* (Scott, Foresman and Co.).

14. S.J. Heine, D.R. Lehman, H.R. Markus et al. (1999), 'Is there a universal need for positive self-regard?', *Psychological Review* 106: pp. 766–94.

15. Eibl-Eibesfeldt (1989), pp. 308–9.
16. Baumeister et al. (2003).
17. *Diagnostic and Statistical Manual of Mental Disorders* (2013) (American Psychiatric Association, fifth edition).
18. M.T. Treadway, N.A. Bossaller, R.C. Shelton et al. (2012), 'Effort-based decision making in major depressive disorder: A translational model of motivational anhedonia', *Journal of Abnormal Psychology* 121: pp. 553–8.
19. D. Hirshleifer and T. Shumway (2003), 'Good day sunshine: Stock returns and the weather', *Journal of Finance* 58: pp. 1009–32.
20. R. Bregman (2017/2018), *Utopia for Realists* (Bloomsbury), pp. 163–5.
21. R.A. Easterlin (1974), 'Does economic growth improve the human lot? Some empirical evidence', in R. David and M. Reder (eds), *Nations and Households in Economic Growth* (Academic Press).
22. A.E. Clark, P. Frijters and M.A. Shields (2008), 'Relative income, happiness, and utility: An explanation for the Easterlin paradox and other puzzles', *Journal of Economic Literature* 46: pp. 95–144.
23. S. Redstone (2001), *A Passion to Win* (Simon & Schuster), chapters 1 and 2.
24. Ford (1922), chapters 1, 2, 6, 13 and 14.
25. Ford (1922), chapters 1, 2, 6, 13 and 14.
26. Haidt (2012/2013).
27. J. Kabat-Zinn (1990), *Full Catastrophe Living* (Dell).

CHAPTER 12

1. https://www.youtube.com/watch?v=5DcdONaKSQM; https://www.youtube.com/watch?v=HAEzfdBMN5k.
2. https://www.youtube.com/watch?v=-_kXlGvB1uU.
3. M.E. Spiro (1979/1996), *Gender and Culture* (Transaction), p. x.
4. A.S. Rossi (1984), 'Gender and parenthood', *American Sociological Review* 49: pp. 1–19.
5. C.H. Sommers (1994/1995), *Who Stole Feminism?* (Touchstone).

6. https://www.youtube.com/watch?v=hg3umXU_qWc. See also: C.N. Adichie (2014), *We Should All Be Feminists* (Fourth Estate).

7. https://www.youtube.com/watch?v=hg3umXU_qWc.

8. Rubenstein and Alcock (2019), p. 346.

9. Rubenstein and Alcock (2019), pp. 288, 297.

10. Eibl-Eibesfeldt (1989), p. 283.

11. Eibl-Eibesfeldt (1989), p. 275.

12. Eibl-Eibesfeldt (1989), p. 268.

13. Eibl-Eibesfeldt (1989), p. 276.

14. D.A. Puts (2010), 'Beauty and the beast: Mechanisms of sexual selection in humans', *Evolution and Human Behavior* 31: pp. 157–75.

15. Rubenstein and Alcock (2019), pp. 291–3.

16. Rubenstein and Alcock (2019), p. 312.

17. Rubenstein and Alcock (2019), pp. 349–50.

18. Rubenstein and Alcock (2019), p. 322.

19. Rubenstein and Alcock (2019), p. 287.

20. Rubenstein and Alcock (2019), p. 215.

21. Rubenstein and Alcock (2019), p. 294.

22. Rubenstein and Alcock (2019), p. 295.

23. Rubenstein and Alcock (2019), p. 296.

24. S. Stewart-Williams and A.G. Thomas (2013), 'The ape that thought it was a peacock: Does evolutionary psychology exaggerate human sex differences?', *Psychological Inquiry* 24: pp. 137–68.

25. Rubenstein and Alcock (2019), p. 163.

26. Rubenstein and Alcock (2019), p. 165.

27. Rubenstein and Alcock (2019), pp. 160–1.

28. A.F. Dixson (1980), 'Androgens and aggressive behavior in primates: A review', *Aggressive Behavior* 6: pp. 37–67.

29. M. Hines, M. Constantinescu and D. Spencer (2015), 'Early androgen exposure and human gender development', *Biology of Sex Differences* 6: p. 3.

30. M. Hines, M. Constantinescu and D. Spencer (2015).

31. S. Bertelloni, G.I. Baroncelli, P. Garofalo et al. (2010). 'Androgen therapy in hypogonadal adolescent males', *Hormone Research in Paediatrics* 74: pp. 292–6.

32. Dixson (1980).

33. T.O. Oyegbile and C.A. Marler (2005), 'Winning fights elevates testosterone levels in California mice and enhances future ability to win fights', *Hormones and Behavior* 48: pp. 259–67.

34. P.C. Bernhardt, J.M. Dabbs Jr, J.A. Fielden et al. (1998), 'Testosterone changes during vicarious experiences of winning and losing among fans at sporting events', *Physiology & Behavior* 65: pp. 59–62.

35. S.N. Geniole, B.M. Bird, J.S. McVittie et al. (2020), 'Is testosterone linked to human aggression? A meta-analytic examination of the relationship between baseline, dynamic, and manipulated testosterone on human aggression', *Hormones and Behavior* 123: p. 104644.

36. Geniole et al. (2020).

37. J.M. Dabbs Jr, R.L. Frady, T.S. Carr et al. (1987), 'Saliva testosterone and criminal violence in young adult prison inmates', *Psychosomatic Medicine* 49: pp. 174–82.

38. For a review of the following material, see: Z.R. Donaldson and L.J. Young (2008), 'Oxytocin, vasopressin, and the neurogenetics of sociality', *Science* 322: pp. 900–4.

39. S.P. Borrow and N.M. Cameron (2012), 'The role of oxytocin in mating and pregnancy', *Hormones and Behavior* 61: pp. 266–76.

40. H. Walum, L. Westberg, S. Henningsson et al. (2008), 'Genetic variation in the vasopressin receptor 1a gene (AVPR1A) associates with pair-bonding behavior in humans', *Proceedings of the National Academy of Sciences* 105: pp. 14153–6.

41. D. Marazziti, S. Baroni, F. Mucci et al. (2019), 'Sex-related differences in plasma oxytocin levels in humans', *Clinical Practice and Epidemiology in Mental Health* 15: pp. 58–63.

42. P.S. Churchland and P. Winkielman (2012), 'Modulating social behavior with oxytocin: How does it work? What does it mean?', *Hormones and Behavior* 61: pp. 392–9.
43. Adichie (2014), p. 29.
44. Brown (1991).
45. Eibl-Eibesfeldt (1989), p. 265.
46. Eibl-Eibesfeldt (1989), p. 610.
47. Eibl-Eibesfeldt (1989), pp. 265–6.
48. Spock (1945/1979), p. 357.

CHAPTER 13

1. https://www.youtube.com/watch?v=mQZmCJUSC6g.
2. https://www.youtube.com/watch?v=MOyvYnkdEcc.
3. A. Dworkin (1993), 'Sexual economics: The terrible truth', *Letters from a War-zone* (Lawrence Hill).
4. N. Friday (1973/1975), *My Secret Garden* (Quartet).
5. Friday (1973/1975), p. 3.
6. Eibl-Eibesfeldt (1989), p. 238.
7. Puts (2010).
8. Stewart-Williams and Thomas (2013).
9. Rubenstein and Alcock (2019), p. 348.
10. S.W. Gangestad and R. Thornhill (2008), 'Human oestrus', *Proceedings of the Royal Society B* 275: pp. 991–1000.
11. Lorenz (1966/2002), p. 100.
12. J. Tweedie, 'Introduction', Friday (1973/1975).
13. G. Greer (1970/2012), *The Female Eunuch* (Fourth Estate).
14. Lorenz (1966/2002), p. 175.
15. E.L. Zurbriggen and M.R. Yost (2004), 'Power, desire, and pleasure in sexual fantasies', *Journal of Sex Research* 41: pp. 288–300.
16. Puts (2010).

17. Friday (1973/1975), p. 209.

18. https://www.youtube.com/watch?v=sjJPnrWSU3Y.

19. I. Allende (1982/1994), *The House of the Spirits* (Black Swan), p. 269.

20. Friday (1973/1975), p. 114.

21. Friday (1973/1975), p. 133.

22. Friday (1973/1975), p. 121.

23. C.A. MacKinnon (1987), 'A feminist/political approach: Pleasure under patriarchy', in J.H. Geer and W.T. O'Donohue (eds), *Theories of Human Sexuality* (Plenum).

24. https://worldpopulationreview.com/country-rankings/rape-statistics-by-country.

25. Puts (2010).

26. Lorenz (1966/2002), p. 124.

27. https://www.youtube.com/watch?v=Q0ELclZ7XeI. https://www.youtube.com/watch?v=axLKkxRhRmg.

28. E.L. Zurbriggen and M.R. Yost. (2004), 'Power, desire, and pleasure in sexual fantasies', *Journal of Sex Research* 41: pp. 288–300.

29. Puts (2010); D.M. Buss (1989), 'Sex differences in human mate preferences: Evolutionary hypotheses tested in 37 cultures', *Behavioral and Brain Sciences* 12: pp. 1–49.

30. H. Wouk (1951), *The Caine Mutiny* (Jonathan Cape), p. 375.

31. S. Agron, C.A. de March, R. Weissgross et al. (2023), 'A chemical signal in human female tears lowers aggression in males', *PLoS Biology* 21: p. e3002442.

32. Mitchell, M. (1936), *Gone with the Wind* (Macmillan), Chapter 54.

33. Friday (1973/1975), p. 27.

34. https://www.bbc.co.uk/news/uk-66982160 (published 4 October 2023); https://www.bbc.co.uk/news/health-66775015 (published 12 September 2023).

35. Lorenz (1966/2002), p. 120.

36. Lorenz (1966/2002), p. 121.

37. Friday (1973/1975), pp. 128–9.

38. R. Kipling (1971), *Short stories: Volume 2* (Penguin), p. 11.

CHAPTER 14

1. Spiro (1979/1996).
2. Spiro (1979/1996), p. 73.
3. Spiro (1979/1996), p. xix.
4. Spiro (1979/1996), pp. xxii, 37.
5. Spiro (1979/1996), p. 41.
6. Spiro (1979/1996), p. xxiv.
7. Spiro (1979/1996), p. 20.
8. Spiro (1979/1996), p. 20.
9. Spiro (1979/1996), p. 19.
10. Spiro (1979/1996), p. xviii.
11. Spiro (1979/1996), p. 31.
12. Spiro (1979/1996), p. 32.
13. Adichie (2014), p. 10.
14. Spiro (1979/1996), p. 27.
15. Spiro (1979/1996), p. 28.
16. Spiro (1979/1996), p. 43.
17. Spiro (1979/1996), p. 24.
18. Spiro (1979/1996), p. 24.
19. Spiro (1979/1996), p. 58.
20. Spiro (1979/1996), p. xviii.
21. Spiro (1979/1996), p. 25.
22. Spiro (1979/1996), p. 25.
23. Spiro (1979/1996), p. xxv.
24. Spiro (1979/1996), p. 60.
25. Spiro (1979/1996), pp. 109–10.
26. Royal Society (2014), *A Picture of the UK Scientific Workforce* (Royal Society).

27. C. Goldin (2014), 'A grand gender convergence: Its last chapter', *American Economic Review* 104: pp. 1091–119.

28. Spiro (1979/1996), p. 106.

29. Eibl-Eibesfeldt (1989), p. 605.

30. Quoted in: Eibl-Eibesfeldt (1989), p. 283.

31. C. Servin-Barthet, M. Martínez-García, C. Pretus et al. (2023), 'The transition to motherhood: Linking hormones, brain and behaviour', *Nature Reviews Neuroscience* 24: pp. 605–19.

32. For a broad review, see: Pinker (2002/2019), Chapter 18.

33. J. Armstrong and J. Ghaboos (2019), *Women Collaborating with Men: Everyday Workplace Inclusion* (Murray Edwards College).

34. J.P. McClure and J.M. Brown (2008), 'Belonging at work', *Human Resource Development International* 11: pp. 3–17.

35. L.E. Gomez and P. Bernet (2019), 'Diversity improves performance and outcomes', *Journal of the National Medical Association* 111: pp. 383–92.

36. Here is one of my favourites, though to enjoy it to the full, you need to understand the rudiments of cricket: https://www.youtube.com/watch?v=SakX1XtRaAU.

37. Eibl-Eibesfeldt (1989), p. 610.

38. Goldin (2014).

39. Goldin (2014), p. 1, 118.

40. https://www.bbc.co.uk/news/world-middle-east-56615521 (published 4 April 2021).

41. S. Knobloch-Westerwick, C.J. Glynn and M. Huge (2013), 'The Matilda effect in science communication: An experiment on gender bias in publication quality perceptions and collaboration interest', *Science Communication* 35: pp. 603–25.

42. https://www.cipd.org/uk/knowledge/evidence-reviews/diversity-inclusion/.

43. C. Napp and T. Breda (2022), 'The stereotype that girls lack talent: A worldwide investigation', *Science Advances* 8: p. eabm3689.

44. S. Kastner (2021), 'Leslie G. Ungerleider', *Neuron* 109: 202–4.
45. https://www.bbc.co.uk/news/world-us-canada-55690001 (published 20 January 2021).
46. Quoted in: Sommers (1994/1995), pp. 256–7.

CHAPTER 15

1. Quoted in: C. Mudde and C. Rovira Kaltwasser (2017), *Populism* (Oxford University Press), p. 17.
2. Charles River Editors (2017), *Athenian Democracy* (CreateSpace Independent Publishing Platform).
3. Mudde and Rovira Kaltwasser (2017), p. 6.
4. Mudde and Rovira Kaltwasser (2017), p. 64.
5. Mudde and Rovira Kaltwasser (2017), p. 45.
6. J. Pakulski (2018), 'The development of elite theory', in H. Best and J. Higley (eds), *The Palgrave Handbook of Political Elites* (Springer).
7. Quoted in: G.M. Gilbert (1947), *Nuremberg Diary* (Signet).
8. Mudde and Rovira Kaltwasser (2017), p. 106.
9. Eibl-Eibesfeldt (1971), p. 163.
10. https://en.wikipedia.org/wiki/Arthur_J._Finkelstein#Quotes; https://www.washingtonpost.com/local/obituaries/arthur-finkelstein-shadowy-campaign-mastermind-and-gop-operative-dies-at-72/2017/08/19/0bd638c6-84e8-11e7-902a-2a9f2d808496_story.html?amp;utm_term=.5b723a2bf7f3&noredirect=on (published 19 August 2017).
11. H. Grassegger (2019), 'The Finkelstein formula', True Story Award 2021, https://truestoryaward.org/story/230.
12. https://www.youtube.com/watch?v=n2Xufahbaq4.
13. Haidt (2012/2013), p. 363.
14. Mudde and Rovira Kaltwasser (2017), p. 116.
15. Lorenz (1966/2002), p. 264.

16. https://www.huffingtonpost.co.uk/entry/jane-goodall-donald-trump-chimps_n_6399a569e4b0c2814648a843 (published 14 December 2022).

17. Mudde and Rovira Kaltwasser (2017), p. 64.

18. J.B. Judis (2016), *The Populist Explosion* (Columbia Global Reports), p. 73.

19. Mudde and Rovira Kaltwasser (2017), p. 66.

20. https://www.bbc.co.uk/news/world-latin-america-67156220 (published 22 October 2023).

21. Mudde and Rovira Kaltwasser (2017), p. 70.

22. Mudde and Rovira Kaltwasser (2017), p. 71.

23. Haidt (2012/2013).

24. Lorenz (1966/2002), pp. 285–6.

25. Haidt (2012/2013), p. 180.

26. Duncan (2010), pp. 200–1.

27. V. Frigo, L. Chen and T. Rogers (2020), 'A cognitive mechanism for the persistence of widespread false beliefs'. PsyArXiv, https://osf.io/preprints/psyarxiv/4zfrp.

28. https://www.bbc.co.uk/news/technology-66472938 (published 13 August 2023).

29. Mudde and Rovira Kaltwasser (2017), p. 111.

Bibliography

Where two dates are given, the first is the original date of publication in English, and the second is the date of the edition listed. Listed editions are those I have used for reference and page numbering in the notes.

Darwin, C. (1871/2013), *The Descent of Man* (Wordsworth Editions)

Eibl-Eibesfeldt, I. (1971), *Love and Hate* (Methuen)

Eibl-Eibesfeldt, I. (1989), *Human Ethology* (Aldine de Gruyter)

Frankl, V. (1959/2004), *Man's Search for Meaning* (Rider)

Friday, N. (1973/1975), *My Secret Garden* (Quartet)

Haidt, J. (2012/2013), *The Righteous Mind* (Penguin)

Lorenz, K. (1966/2002), *On Aggression* (Routledge)

Mudde, C. and Rovira Kaltwasser, C. (2017), *Populism* (Oxford University Press)

Pinker, S. (2002/2019), *The Blank Slate* (Penguin)

Rubenstein, D.R. and Alcock, J. (2019), *Animal Behavior: International Eleventh Edition* (Oxford University Press)

Spiro, M.E. (1979/1996), *Gender and Culture* (Transaction)

Wacks, R. (2006/2014), *Philosophy of Law* (Oxford University Press)

Index

abstract idea machine *see* idea machine
abstraction 24
  moral rules 210
  thoughts 79, 80
  violence as 125
academic criminology 228–9
academic women's studies 262
acceptance, understanding and 184–6
Achebe, Chinua 157, 355
Adams, John 239, 243
Adichie, Chimamanda Ngozi 28, 264–5, 268
  We Should All Be Feminists 264–5, 281, 325
aggression 18–19, 20, 42, 241–3
  in animal groups 18–19, 134, 272, 273
  in children 20
  controlling 201
  defusing 187–8
  discharge of 49
  displays of 42
  instincts and 19
  in mating pairs 134
  redirected 50–1, 134–5
  sexuality and 177–8
  testosterone and 101, 277
  see also Lorenz, Konrad
Agnew, Jonathan 147
agreements 219–21
algorithms 366
all-knowing God 203
Allende, Isabel 301
Allport, Gordon 186–7
alternatives, weight of 47–8
altruism 167
America
  Declaration of Independence 90, 216
  First Amendment, Constitution 221
  free speech 221
  individualism 246–7
  land of opportunity 90
  promoting self-esteem 247–8
  self-esteem scale 246
  striving 246–7

  Task Force to Promote Self-Esteem and Personal and Social Responsibility 248
amygdala 99, 102, 113, 306, 335
Ancient Evenings (Mailer) 254
anger, challenge and 241–2
animal behaviour 4, 16
  hereditary 38
  inconsistency 20–1
  learning in 38, 39
  Lorenz 7–8
  observing 33–4
  order 35
  see also ethology
animal rights 172
animals/animal groups 130–2
  aggression 18–19, 134
  bonding 134–8
  defeating the enemy *see* enemies
  defending territory 134–8, 339
  differentiating males and females 269–70, 271–3
  genes 271, 272
  male aggression 272, 273
  male displays of strength 272–3
  mating practices 275
  mating season 274–5
  mobbing 146
  pairs 134
  parental care 132
  rank displays 339
  reproduction 271–3
  selecting mates 271–3
  submission of partners 314
  testosterone 274–6
  triumph ceremonies 135–8
anterior hypothalamus 100
Anthills of the Savannah (Achebe) 355
Anthony, Susan B. 264, 343
apes 141, 268, 269
Apollo (temple) 369
appeasement gestures 66
Apted, Michael 285–6
Aquinas, Thomas 4, 217, 219

411

INDEX

reasoning *see* reasoning
submission of partners 314–15
testosterone 101, 274–7
see also human groups; men; women
Hume, David 16–17, 206–7, 218, 226, 375
reasoning 16, 206, 232, 375
A Treatise of Human Nature 16
humour 147, 365
Hungary 359–60
hunter-gatherer groups 139, 154–5, 159
Huxley, Julian 40, 44
hypothalamus 100, 101, 102, 275, 335
see also anterior hypothalamus;
ventromedial hypothalamus

idea machine 79, 80, 81, 82, 94, 112
ideas
abstract 53, 79, 82, 126
arbitrary 121–3
competing 112–13, 378
evolution of 117–20
generation of 374
good and bad 80–2
over-simplified, Spock 14, 21, 23, 263, 264,
374
simple 78–80, 89–91, 110, 126, 374
social 118–19
spread of 121–3
The Idiot (book) 209–10
The Idiot (film) 209
immigrants 184
immigration crises 357
imprinting 154
individualism 246–7
inequality
in leadership positions 331–2
in salaries 331–2
see also equality
informed advice 370
informed democracy 371
initiation ceremonies/rites 193, 244
innate behaviour 51
innate releasing mechanisms *see* IRMs (innate
releasing mechanisms)
instincts 11, 14, 18
aggression 19
competing 12
complex 11, 69
culture and 69
human 11, 14, 18, 19, 23–4
learning and 72–3
reason and 9–10, 12
social 12
Spock's perspective 18
intelligence tests 75–8, 106–7
intimacy 183
IQ distributions 266–7

IRMs (innate releasing mechanisms) 36,
382–3
affiliative and aggressive 185
analogy to computer code 35, 37
animal experimentation 38
artificial stimuli 56
authority ranking 160–4, 169
bees' nervous systems 36
birds 43–4
catching prey 43
communal sharing 158–60, 169
competing 44–7, 294, 364–6,
373–4, 375
competing control codes 45
complexity of 373–4, 375
conflict 44–5, 65–7
cormorant colony 37
discharge 48–51
in *Drosophila* 97–8
enemies 176
equality matching 164–7, 169
fluctuations 67–8
hormones 100–1
humans 55–7, 102–4, 287–8, 373–4
instinctive behaviour 37–8, 39
mammals 98–100
market pricing 168, 169
medial preoptic nucleus 100
neural mechanisms 36
neurophysiology of 94
parental commitment and care 328–9
periaqueductal grey 99–100
pleasure and pain 226
purpose of 36
termites' nervous systems 36
isms 286–8
Israel
Finkelstein's election campaign 359
see also kibbutz movement
It's Great to Be Here (show) 310

jackdaws 162
James, E.L. 297, 308
James, William
Principles of Psychology 47, 244
Jane and the Dragon (Bayton) 282–3
Japan
collective ideals 246–7
effort (*doryoku*) 246
endurance (*gaman*) 246
perseverance (*gambari*) 246
striving 246–7
Japanese macaques 339
Japanese quails 275
Jarecke, Kenneth 181
Jerusalem 121
Jesus, teachings of 20, 22

417